THE RACIAL POLICIES OF AMERICAN INDUSTRY

REPORT NO. 26

THE NEGRO IN THE FARM EQUIPMENT AND CONSTRUCTION MACHINERY INDUSTRIES

by

ROBERT OZANNE
Professor of Economics
University of Wisconsin

with the assistance of
MISS ELSA KLEMP

Published by
INDUSTRIAL RESEARCH UNIT, DEPARTMENT OF INDUSTRY
Wharton School of Finance and Commerce
University of Pennsylvania

Distributed by
University of Pennsylvania Press
Philadelphia, Pennsylvania 19104

Library of Congress Catalog Card Number 77-186541
MANUFACTURED IN THE UNITED STATES OF AMERICA
ISBN: 0-8122-9070-4

FOREWORD

In September 1966, the Ford Foundation announced a major grant to the Industrial Research Unit of the Wharton School to fund studies of the Racial Policies of American Industry. The purpose of the research effort, now in its sixth year, is to determine why some industries are more hospitable to the employment of Negroes than are others and why some companies within the same industry have vastly different racial employment policies, and to propose appropriate policy.

The studies have proceeded on an industry by industry basis, under the direction of the undersigned, with Dr. Richard L. Rowan, Associate Professor of Industry, as Associate Director. This study of the farm equipment and construction machinery industries is the twenty-sixth in a series of reports dealing with specific industries. A list of industry studies already published or in process is on the back cover. In addition to the individual industry reports, major studies combining and comparing the findings of the various industry reports have been undertaken. Five of these volumes—*Negro Employment in Basic Industry, Negro Employment in Finance, Negro Employment in Public Utilities, Negro Employment in Southern Industry,* and *Negro Employment in Land and Air Transport*—have recently been published. *Negro Employment in Retail Trade* will be published later in 1972. This study, *The Negro in the Farm Equipment and Construction Machinery Industries,* is one of a group of selected, specialized manufacturing industries which have been given detailed attention in the Racial Policies series.

This study covers two related industries in which Negro employment has deep historical roots, but is not a large proportion of the total work force. It is authored by Professor Robert Ozanne of the University of Wisconsin, who has previously published two distinguished works dealing with the history of labor relations at International Harvester Company, the largest company in these industries. One, *A Century of Labor Management Relations at McCormick and International Harvester* (University of Wisconsin Press, 1967) received the Thomas Newcomen Award in Business History.

Numerous persons contributed to this study which was substantially reorganized and refined by the Industrial Research Unit staff. Miss Elsa Klemp spent many hours developing statistical data and Mrs. Veronica M. Kent typed the manuscript. Mrs. Marie R. Keeney edited the manuscript and constructed the index. Many industrial, union, and government officials also aided the research with data and interviews.

As in most previous reports, the data cited as "in the author's possession" have been carefully authenticated and are on file in the Industrial Research Unit library.

HERBERT R. NORTHRUP, *Director*
Industrial Research Unit
Wharton School of Finance and Commerce
University of Pennsylvania

Philadelphia

March 1972

TABLE OF CONTENTS

LIST OF TABLES

TABLE PAGE

LIST OF FIGURES

FIGURE PAGE

CHAPTER I

Introduction

Conflicting views of the progress in eliminating job discrimination toward Negroes are found in reports by two federal government agencies. One, a study by the U.S. Equal Employment Opportunity Commission, concludes that as "greater progress for Negroes brings forth progressively stronger discrimination against them," and as more Negroes succeed in getting jobs, especially higher-paying jobs, and better education, "fewer of them can expect promotions" and "the bias against them will be stronger."[1] This prediction is contradicted by a Bureau of Labor Statistics survey summarized as follows: "The past decade has seen a marked improvement in the number and kinds of jobs held by black Americans. Some of this occupational upgrading occurred between 1957 and 1962, but the trend accelerated sharply in the 1962-67 period, as nearly 1.1 million Negroes moved into jobs offering higher pay and status."[2] The Bureau of Labor Statistics' report gives as the causes of such improved job opportunities for Negroes the following: "Major factors contributing to these gains included an exceptional increase in the total number of job opportunities provided by the rapid economic expansion of most of the decade, new and expanded programs for retraining . . ., higher levels of educational attainment, and the decline in discrimination in hiring."[3]

Both reports agree that there is still a high level of discrimination against the Negro. The Equal Employment Opportunity Commission puts it thusly: "Two-thirds of the Negroes, Spanish-speaking Americans, and Indians are held back from jobs by outright racial discrimination and not by lack of skills or education,"[4] and the Bureau declares "despite these recent substantial

---

[1] U.S. Equal Employment Opportunity Commission, press release issued in 1968 with publication of Report No. 1 by the Commission.

[2] Claire C. Hodge, "The Negro Job Situation: Has It Improved?" *Monthly Labor Review*, Vol. 92 (January 1969), p. 20.

[3] *Ibid.*

[4] U.S. Equal Employment Opportunity Commission, *loc. cit.*

gains, however, the Negro today still holds a disproportionate share of jobs at the lower end of the occupational ladder and is underrepresented in the higher-skilled, better-paying jobs."[5] Hopefully, the Racial Policies studies of specific industries can give some insight into these conflicting reports and an understanding of the nature of the barriers which still bar Negroes from equal job opportunity in some industries and provide them opportunities in others.

In this study, we examine two closely related manufacturing industries for their problems and progress in equal employment opportunity—farm equipment (machinery) and construction machinery. The production processes of these two industries are quite similar, both are concentrated in the Midwest, and several of the leading farm equipment firms also produce construction machinery, although Caterpillar, the largest construction machinery concern, which was founded to manufacture agricultural implements, is no longer in that business.

As in several previous studies of this series, emphasis is on the major companies. Personal interviews were made with officials of these concerns as well as with civil rights and local and national union leaders. The bulk of the fieldwork was conducted in the 1967-1969 period, but more recent contacts have been made, particularly to assess the impact of layoffs. The concentration on large firm policy may mean that we are dealing with a group more favorable to black employment progress than is the average firm in these industries. Being more visible, more thoroughly and more often inspected by government investigators, and having more resources to devote to equal employment opportunity, larger companies often—but not always by any means—have a better record in such matters. Data from the Equal Employment Opportunity Commission for 1969, which cover all firms with 100 or more employees, provide a measure by which our large company figures can be evaluated in terms of industry norms.

Chapter II of this report describes the industries under study with special attention to the factors affecting the extent and character of Negro employment. After a survey of the trend of Negro employment before 1960 in Chapter III, the balance of the study is devoted to a discussion of Negro employment and the factors affecting it in the period since 1960.

[5] Hodge, *loc. cit.*

CHAPTER II

The Farm Equipment and Construction Machinery Industries

The farm equipment (or implement) and construction machinery industries derived from the common desire of man, first on the farm, then in applying the same concepts in building roads, offices, factories, and houses, to create machinery which would handle heavy tasks, or perform them or other functions better or easier. The modern farm machinery industry developed first. It had its genesis in western Virginia where in 1832 Cyrus Hall McCormick built and operated a horse-powered reaper.[6] In 1847, McCormick moved to Chicago and built his own factory. This plant was by 1902 producing 35 percent of America's harvesting machinery. In that year, the McCormick family company merged with four others to form International Harvester Company, a giant trust producing 85 percent of the country's harvesting machinery. In 1914, a federal court found that the agricultural implement trust violated the Sherman Antitrust Act and ordered it dissolved.[7] International Harvester retained the McCormick plant and others and continues to be the largest corporation in the industry, but in terms of farm machinery, Deere and Company has now passed it in sales.

Construction machinery is a newer industry which has developed enormously since World War II. Caterpillar, the leading producer, was founded in 1890 in California as a farm tractor producer, but became famous for the application of crawler tracks to replace wheels. The company acquired the Peoria, Illinois, lo-

[6] The historical material relating to International Harvester is based on Robert Ozanne, *A Century of Labor-Management Relations at McCormick and International Harvester* (Madison: The University of Wisconsin Press, 1967).

[7] *United States* v. *International Harvester Company,* 214 Fed. 987 (D.C., D. Minn., 1914). This decision was later affirmed by the Supreme Court in an application for further remedies against the alleged "trust." 274 U.S. 693 (1927).

cation which is now its headquarters in 1909, and sold its farm equipment business to Deere in 1935. This is in contrast to the experience of several farm equipment companies which have added construction machinery—especially road or earth moving lines—in recent years. Much of this construction equipment is in the form of attachments to tractors for earth moving purposes.

DEFINITION OF THE INDUSTRIES

Table 1 reproduces the official United States Standard Industrial Classification (SIC) numbers and definitions of the two industries. Our data will be largely three digit classification material, and therefore for construction machinery especially, will include mining machinery and other related industrial groups. It is not believed, however, that the inclusion of these smaller subgroups in our data will seriously affect the results, especially since the manufacturing processes for the entire construction and related machinery group are quite similar.

INDUSTRIAL STRUCTURE

Table 2 lists pertinent statistical information for fourteen large companies in the farm and construction machinery industries. International Harvester is by far the largest in terms of both sales and employment with Caterpillar second. Deere and Company, the third largest, sells more farm implements than any other concern, Harvester's sales being bolstered by its sizable truck and diesel engine sales.

The list of companies in Table 2 is necessarily somewhat arbitrary, both as to inclusions and exclusions. White Motor, like International Harvester, is also a leading truck manufacturer. Since the mid-1950's, however, it has acquired two major farm equipment concerns—Oliver and Minneapolis-Moline—and is therefore included. Allis-Chalmers, a major manufacturer of heavy electrical equipment and other products, is included because of its major position in both the construction and farm machinery industries. Clark Equipment and Ingersoll-Rand are included although both produce products other than construction machinery.

Table 2 excludes several key concerns for which data are either unavailable or for which farm or construction machinery products are a minor segment of total sales. Thus J. I. Case, one of

TABLE 1. *Farm Equipment and Construction Machinery Industries, Standard Industrial Classification Manual Definitions*

Farm Machinery and Equipment
(Group No. 352, Industry No. 3522)

Agricultural implements and machinery
Ammonia applicators and attachments (agricultural machinery)
Axles, farm tractor
Balers: hay, cotton, etc.
Barn stanchions and standards
Blowers, forage
Blowers and cutters, ensilage
Brooders
Calf savers (farm equipment)
Cattle feeding, handling, and watering equipment
Cattle oilers (farm equipment)
Chicken brooders
Cleaning machines for fruits, grains and vegetables
Clippers, hair: for animal use—hand and electric
Combines (harvester-threshers)
Conveyors, farm (agricultural machinery)
Corn huskers, shellers, and shredders
Cotton pickers and strippers (operational machinery)
Cream separators (agricultural equipment)
Crop driers
Crushers, feed (agricultural machinery)
Cultivators (agricultural implements)
Curers, tobacco
Cutters, ensilage
Drags and stone boats (agricultural equipment)
Driers: grain, hay, and seed (agricultural implements)
Dusters, mechanical: agricultural
Elevators, farm
Farm machinery and equipment
Farm tractors
Feed grinders, crushers, and mixers (agricultural machinery)
Feeders, chicken
Fertilizing machinery (agricultural machinery)
Field type rotary tillers (agricultural machinery)
Fruit grading, cleaning, and sorting machines
Gates, holding (farm equipment)
Grading, cleaning, and sorting machines: fruit, grain, and vegetable
Grain drills, including legume planters (agricultural machinery)
Grain grading, cleaning, and sorting machines
Grass catchers, lawnmower
Grain stackers
Grinders and crushers, feed (agricultural machinery)
Hair clippers for animal use, hand and electric
Hammer and roughage mills (agricultural machinery)
Harrows
Harvesting machines
Haying machines: mowers, rakes, loaders, stackers, balers, presses, etc.
Hedge trimmers, electric
Hog feeding, handling, and watering equipment
Hulling machinery, agricultural
Huskers, corn
Incubators, except laboratory and infant
Land rollers and levelers (agricultural machinery)
Lawn mowers, hand and power
Listers
Loaders, farm type (general utility)
Milking machines
Mowers, hay and lawn; hand and power
Mowers, highway: self-propelled
Peanut combines, diggers, packers and threshers (agricultural equipment)
Planting machines, agricultural
Plows, agricultural
Potato diggers, harvesters, and planters (agricultural machinery)

TABLE 1—Continued

Poultry brooders, feeders, and waterers
Poultry vision control devices
Presses and balers: hay, cotton, etc.
Rakes, hay (agricultural machinery)
Reapers
Rotary hoes (agricultural machinery)
Roughage mills (agricultural machinery)
Seeders (agricultural machinery)
Separators, grain and berry (agriculture machinery)
Shears, sheep: power
Shellers, nut (agricultural machinery)
Shredders (agricultural machinery)
Silo fillers (agricultural machinery)
Soil pulverizers and packers (agricultural machinery)
Sorting machines for fruits, grains, vegetables, and other agricultural products
Sprayers, hand: agricultural
Spraying machines (agricultural machinery)
Spreaders, fertilizer
Stone boats (agricultural equipment)
Tedders, hay
Threshing machines, agricultural
Tobacco curers
Towers, windmill
Tractors: wheel, except contractors, off-highway types
Transmissions, farm tractors
Transplanters
Volume guns (irrigation equipment)
Water troughs
Weeding machines, agricultural
Wheels, for farm machinery
Windrowers

Construction Machinery and Equipment
(Group No. 353, Industry No. 3531)

Aggregate spreaders
Airport construction machinery
Asphalt plants, including travel-mix type
Backfillers, self-propelled
Backhoes
Ballast distributors
Batching plants, bituminous
Batching plants, for aggregate concrete and bulk cement
Blades for graders, scrapers, dozers, and snow plows
Breakers, paving
Buckets: clamshell, concrete, dragline, drag scraper, orange peel, shovel, and hoe dipper
Bulldozers (construction machinery)
Capstans, ship
Carriers, crane
Chip spreaders, self-propelled
Cleaners, catch basin
Compactors, soil: vibratory-pan and vibratory-roller types
Concrete buggies, powered
Concrete grouting equipment
Concrete gunning equipment
Concrete plants
Construction machinery, except oil field machinery, and tools: dredging, excavating road and airport construction, etc.
Cranes, except industrial plant
Cranes, ship
Crushers, portable
Derricks, except oil and gas field
Derricks, ship
Distributors (construction machinery)
Ditchers, ladder: vertical boom or wheel
Dozers, tractor mounted, material moving
Draglines, powered
Drags, road (construction and road maintenance equipment)
Dredging machinery
Entrenching machines
Excavators: cable, clamshell, crane, derrick, dragline, power shovel, etc.
Extractors, piling

TABLE 1—Continued

Finishers, concrete and bituminous: powered
Finishers and spreaders (construction equipment)
Grader attachments, elevating
Graders, road (construction machinery)
Grapples: rock, wood, etc.
Grinders, stone: portable
Hammer mills (rock and ore crushing machines), portable
Hammers, pile driving
Line markers, self-propelled
Loaders, tractor shovel: self-propelled
Locomotive cranes
Logging equipment
Mixers: concrete, ore, sand, slag, plaster, mortar, and bituminous
Mortar mixers
Mud jacks
Pavers
Planers, bituminous
Pile drivers (construction machinery)
Plaster mixers
Plows, earth: heavy duty
Power cranes, draglines, and shovels
Pulverizers, stone: portable
Railway track equipment: rail layers, ballast distributors, etc.
Rakes, land clearing: mechanical
Road construction and maintenance machinery
Rock crushing machinery, portable
Rollers, road: steam or other power
Rollers, sheepsfoot and vibratory
Sand mixers
Scarifiers, road
Scrapers (construction machinery)
Screeds and screeding machines
Screeners, portable
Ship cranes and derricks
Ship winches
Shovel loaders, wheel tractor
Shovels, power
Silos, cement (batch plant)
Slag mixers
Snow plow attachments
Spreaders and finishers (construction equipment)
Subgraders, construction equipment
Subsoiler attachments, tractor mounted
Surfacers, concrete grinding
Tampers, powered
Tamping equipment, rail
Teeth, bucket and scarifier
Tractors, crawler
Tractors, contractors' off highway
Tractors, tracklaying
Trucks, off-highway: heavy duty motor
Vibrators for concrete construction
Winches, all types

Source: U.S. Bureau of the Budget, *Standard Industrial Classification Manual, 1967* (Washington: Government Printing Office, 1967), pp. 151-152.

Note: Group 353 also includes Industries 3532-3537: mining and materials handling machinery and equipment.

TABLE 2. *Fourteen of the Largest Farm Equipment and Construction Machinery Companies 1970 Statistics*

Company and 1970 Rank among Industrial Corporations		Headquarters	Sales ($000)	Assets ($000)	Net Income ($000)	Number of Employees	Net Income as a Percent of Sales
International Harvester	(32)	Chicago	2,711,535	2,217,571	52,432	101,920	1.9
Caterpillar Tractor	(42)	Peoria, Ill.	2,127,800	1,813,600	143,800	64,040	6.8
Deere	(103)	Moline, Ill.	1,137,695	1,509,692	45,995	41,734	4.0
Massey-Ferguson [a]		Toronto, Canada	937,903	1,011,719	(19,688) [b]	47,386	—
Allis-Chalmers Manufacturing	(139)	West Allis, Wis.	870,076	768,909	15,022	28,158	1.7
White Motor	(150)	Cleveland	810,009	618,329	(19,340) [b]	15,936	—
Ingersoll-Rand	(157)	New York	765,818	741,661	65,469	31,548	8.5
Clark Equipment	(178)	Buchanan, Mich.	671,007	501,938	35,281	22,100	5.3
Koehring [c]	(405)	Milwaukee	237,069	208,132	4,762	7,300	2.0
American Hoist & Derrick [c]	(452)	St. Paul	196,673	162,057	2,135	6,034	1.1
Gardner-Denver	(456)	Quincy, Ill.	193,943	177,741	23,258	6,355	12.0
Chicago Pneumatic Tool	(526)	New York	156,619	171,017	14,191	9,700	9.1
Harnischfeger	(543)	Milwaukee	149,960	143,483	1,833	5,058	1.2
Bucyrus-Erie	(548)	South Milwaukee	148,422	136,515	10,344	5,007	7.0

Source: For the eleven largest, except for Massey-Ferguson, *Fortune*, Vol. LXXXIII (May 1971), pp. 172-187; for Numbers 12-14, *Fortune*, Vol. LXXXIII (June 1971), pp. 102-103; and for Massey-Ferguson, *Fortune*, Vol. LXXXIV (August 1971), pp. 152-153.

[a] In 1970, *Fortune* ranked Massey-Ferguson (previously Massey-Harris) 82nd in size among industrial corporations headquartered outside of the United States.

[b] Loss.

[c] Fiscal year November 30, 1970.

the top seven manufacturers of farm equipment, and also a manufacturer of construction machinery, is excluded because it is now a division of Tenneco. Ford Motor Company, also one of the major seven farm machinery producers by virtue of its tractor division, is also excluded. Other major concerns which have significant farm machinery divisions include Sperry-Rand and AVCO.

Construction machinery manufacturing is even less dominated by a few concerns than is farm machinery. While all farm machinery companies make construction equipment—if only tractor attachments—Caterpillar and the smallest eight companies listed in Table 2 are solely in the construction machinery industry. Other construction machinery concerns tend to be small, usually with single plants, or divisions of larger companies. Foremost among the latter are General Motors, a major factor in the business, and Warner & Swasey Gradall Division.

The lack of industrial concentration, especially in construction machinery, tends to make it difficult to discuss racial policies for the industry as a whole. Our basic data, however, will include several of the smaller concerns, and for 1969, the two industries as a whole. We have excluded from our study any analysis of policies of companies such as Ford or General Motors, which were discussed in other studies.[8]

Production Processes and Work Character

Farm and construction machinery concerns, like automobile manufacturers, operate production systems that are product-integrated to varying degrees. International Harvester owns a steel making concern and makes its own engines. Caterpillar and Deere manufacture engines, but not steel. Others buy diesel engines from these concerns, or from Cummins Engines or General Motors. Similarly, their parts making capacities vary from completely integrated systems to ones that depend on a host of sup-

[8] See Herbert R. Northrup, *et al.*, *Negro Employment in Basic Industry*, Studies of Negro Emploment, Vol. I (Philadelphia: Industrial Research Unit, Wharton School of Finance and Commerce, University of Pennsylvania, 1970), Part Two, for an analysis of Ford and General Motors policies. Despite its significance in electrical manufacturing, Allis-Chalmers policies were not analyzed in the study of that industry, but are examined herein. See Theodore V. Purcell and Daniel P. Mulvey, *The Negro in the Electrical Manufacturing Industry*, The Racial Policies of American Industry, Report No. 27 (Philadelphia: Industrial Research Unit, Wharton School of Finance and Commerce, University of Pennsylvania, 1971).

pliers. Parts manufacturing involves the typical metal working equipment familiar to factory operations in many businesses—cutting, stamping, boring, finishing, etc. Some of this work requires considerable skill; much is only semiskilled and requires little training. In addition, the larger firms do a considerable amount of casting and molding in captive foundries, often areas where, because of the hot, dirty character of the work, Negroes were first employed.

Farm and construction equipment is usually assembled on lines which superficially resemble those in the automobile industry. As in the automobile industry, the machine is put together on the line, but the line moves at a much slower pace. Output per day is considerably less than in an automobile plant where 50 or more cars may be produced in an hour, but somewhat more than in the aircraft segment of the aerospace industry where one completed product per day is likely to be a very high production rate. The reason for this rate of production is, of course, a function of the considerably less output and less standardization in farm and construction equipment than in automobile production, but a simpler product in many ways than aircraft. Farm and construction machinery usually involves quite different accessories for different utilization purposes and more complicated accessories and controls than in automobile manufacturing. In terms of employment this implies that the farm equipment or construction machinery assembly line worker must do a greater variety of operations and therefore possess more knowledge and skill than his automobile counterpart, although somewhat less of both than the aircraft assembly worker.

MANPOWER

Employment in the two industries is shown in Table 3 for the period since 1958. In this period, the number of employees in farm equipment grew from 112,000 to a 1967 peak of 142,200, and then declined sharply to 128,700 by 1970. On the other hand, construction machinery employment has both expanded more rapidly, and declined less in the recent industrial cutbacks, growing from 205,300 to a 1969 peak of 294,100, before dropping to 290,300 in 1970. At the present rates of growth and change in the two industries, construction machinery, which employed less than twice the number of employees in 1958 as did farm machinery, could employ three times the number of the latter before the end of the 1970's.

TABLE 3. *Farm Equipment and Construction Machinery Industries*
Total and Production Worker Employment, 1958-1970

Year	Combined Industries			Farm Equipment			Construction Machinery		
	All Employees	Production Workers	Percent Production Workers	All Employees	Production Workers	Percent Production Workers	All Employees	Production Workers	Percent Production Workers
	(000)			(000)			(000)		
1958	318.0	211.8	66.6	112.7	81.5	72.3	205.3	130.3	63.5
1959	347.9	237.6	68.3	122.3	88.8	72.6	225.6	148.8	66.0
1960	333.6	223.9	67.1	112.3	78.4	69.8	221.3	145.5	65.7
1961	309.7	206.1	66.5	108.8	76.2	70.0	200.9	129.9	64.7
1962	323.7	220.3	68.1	112.3	80.4	71.6	211.4	139.9	66.2
1963	337.6	232.1	68.8	119.6	86.6	72.4	218.0	145.5	66.7
1964	362.8	252.6	69.6	126.5	92.1	72.8	236.3	160.5	67.9
1965	391.9	274.6	70.1	135.7	99.0	73.0	256.2	175.6	68.5
1966	426.0	300.0	70.4	148.4	109.9	74.1	277.6	190.1	68.5
1967	424.7	293.0	69.0	149.2	109.1	73.1	275.5	183.9	66.8
1968	423.1	287.9	68.0	142.6	102.0	71.5	280.5	185.9	66.3
1969	429.5	293.1	68.2	135.4	96.1	71.0	294.1	197.0	67.0
1970	419.0	282.4	67.4	128.7	91.1	70.8	290.3	191.3	65.9

Source: U.S. Bureau of Labor Statistics, *Employment and Earnings, United States, 1909-70,* **Bulletin 1312-7, pp. 159-161;** and *Employment and Earnings,* **Vol. 17 (March 1971)**, Table B-2.

The declines in employment in both industries in 1961 and 1970 show that cyclical variations are significant. To Negroes this is an important factor because in many plants Negroes comprise a disproportionate number of the most recently hired, and therefore, under seniority provisions, the most vulnerable to layoffs when employment declines.

Occupational Distribution

The farm equipment and construction machinery industries, unlike many others, remain primarily production worker, or blue collar, in terms of their employment. Table 3 shows that about two-thirds of the employees are production workers. This continued blue collar orientation is explored in more detail in Table 4, which shows employment distribution by broad occupational groups for the two industries combined, and for each separately for establishments reporting to the Equal Employment Opportunity Commission in 1969.

The data in Table 4 confirm the one-third white collar, two-thirds blue collar division indicated by Table 3 figures. Construction machinery has a slightly higher white collar ratio, but the differences do not appear significant. The white collar breakdown reveals a relatively low professional and technical employee force (less than 10 percent of all employees). This is similar to the situation in the automobile and rubber tire industries, but quite different from industries such as aerospace, petroleum, chemicals or electrical manufacturing, in which 17 to 30 percent of all employees are professionals or technicians.[9] In view of the paucity of Negro technicians and trained professionals, the relative modesty of farm and construction needs for such employees is favorable for black employment.

By far the largest percentage of employees in both industries are classified as operatives. Such employees include machine tenders and operatives, assembly line employees and related workers. The jobs performed by operatives vary in skill from entry jobs which require little training or background to those which require considerable experience and training and which verge upon craft work in skill requirements.

Craftsmen make up the second largest occupational group. In construction machinery, they are one-fifth of the labor force, while in farm equipment, they are only 13.2 percent and opera-

[9] See Northrup, *et al.*, *op. cit.*, p. 724; and Purcell and Mulvey, *op. cit.*, p. 16.

TABLE 4. *Farm Equipment and Construction Machinery Industries*
Distribution of Employment by Occupational Group
864 Establishments
United States, 1969

Occupational Group	Combined Industries		Farm Equipment		Construction Machinery	
	All Employees	Percent of Total	All Employees	Percent of Total	All Employees	Percent of Total
Officials and managers	29,271	8.6	9,857	9.3	19,414	8.3
Professionals	15,981	4.7	4,604	4.3	11,377	4.8
Technicians	17,213	5.0	4,946	4.6	12,267	5.2
Sales workers	8,225	2.4	1,910	1.8	6,315	2.7
Office and clerical	41,099	12.0	12,198	11.4	28,901	12.3
Total white collar	111,789	32.7	33,515	31.4	78,274	33.3
Craftsmen	71,197	20.8	14,121	13.2	57,076	24.2
Operatives	119,320	34.9	41,001	38.4	78,319	33.3
Laborers	30,489	8.9	13,282	12.5	17,207	7.3
Service workers	9,216	2.7	4,788	4.5	4,428	1.9
Total blue collar	230,222	67.3	73,192	68.6	157,030	66.7
Total	342,011	100.0	106,707	100.0	235,304	100.0

Source: U.S. Equal Employment Opportunity Commission, 1969.

Note: Data represent 229 farm machinery and 635 construction machinery establishments.

tives outnumber craftsmen nearly three to one. The difference in the occupational distribution between the two industries is most likely the result of a great variety of products and small production runs in the construction machinery industry which do not permit the "breakdown" of jobs and the use of less skilled personnel in many operations. Farm equipment, on the other hand, has larger facilities on average which use more laboring and service workers. These differences in blue collar distribution should be modified by indications from field work that some plants in the construction machinery industry have classified as craftsmen employees who in other industries are considered operatives.

Female Employment

Table 5 shows the number and proportion of female employees in the two industries for the period since 1958. Females have consistently comprised 8 to 10 percent of the work force. Nearly all of these females are found in the office and clerical group; there is a much smaller concentration in semiskilled and unskilled blue collar work. Obviously these industries, with their needs for heavy factory work, have not been hospitable to female employment, white or black, except in traditional office jobs.

Earnings and Hours

Workers in both industries are well paid, receiving considerably above the average manufacturing and average durable goods hourly and weekly wages, as Table 6 shows. The major companies, International Harvester, Caterpillar, and Deere, as well as others, often follow the automobile industry pattern and frequently pay above that rate as well. Other companies may follow the trend established by major steel concerns. In either case, not only wage, but fringe benefits as well are generally high, and working conditions comparably favorable. As a result, companies in these industries generally have little trouble attracting workers and therefore jobseekers usually have considerable competition for employment.

Unionization

Since the late 1930's most of the companies in these two industries have been unionized. The dominant organization is the United Automobile, Aerospace and Agricultural Implement Work-

TABLE 5. *Farm Equipment and Construction Machinery Industries Total and Female Employment 1958-1970*

Year	Combined Industries			Farm Equipment			Construction Machinery		
	All Employees (000)	Female Employees (000)	Percent Female	All Employees (000)	Female Employees (000)	Percent Female	All Employees (000)	Female Employees (000)	Percent Female
1958	318.0	30.8	9.7	112.7	10.5	9.3	205.3	20.3	9.9
1959	347.9	31.6	9.1	122.3	10.8	8.8	225.6	20.8	9.2
1960	333.6	30.5	9.1	112.3	10.2	9.1	221.3	20.3	9.2
1961	309.7	28.3	9.1	108.8	9.6	8.8	200.9	18.7	9.3
1962	323.7	28.2	8.7	112.3	9.5	8.5	211.4	18.7	8.8
1963	337.6	28.9	8.6	119.6	9.9	8.3	218.0	19.0	8.7
1964	362.8	30.1	8.3	126.5	10.3	8.1	236.3	19.8	8.4
1965	391.9	32.0	8.2	135.7	11.1	8.2	256.2	20.9	8.2
1966	426.0	36.0	8.5	148.4	12.7	8.6	277.6	23.3	8.4
1967	424.7	38.2	9.0	149.2	13.7	9.2	275.5	24.5	8.9
1968	423.1	38.8	9.2	142.6	13.4	9.4	280.5	25.4	9.1
1969	429.5	40.1	9.3	135.4	13.7	10.1	294.1	26.4	9.0
1970	419.0	40.4	9.6	128.7	13.2	10.3	290.3	27.2	9.4

Source: U.S. Bureau of Labor Statistics, *Employment and Earnings, United States, 1909-70*, Bulletin 1312-7; and *Employment and Earnings*, Vol. 17 (March 1971), Tables B-2 and B-3.

TABLE 6. *Farm Equipment and Construction Machinery Industries and All Manufacturing Industries Average Earnings and Hours Worked Production Workers, 1970*

Industry Group	Average Weekly Earnings	Average Hourly Earnings	Average Weekly Hours
All manufacturing	$133.73	$3.36	39.8
Durable goods	143.47	3.56	40.3
Nondurable goods	120.43	3.08	39.1
Total farm and construction machinery	150.68	3.72	40.4
Farm machinery and equipment	147.66	3.71	39.8
Construction and related machinery	153.71	3.74	41.1

Source: U.S. Bureau of Labor Statistics, *Employment and Earnings*, Vol. 17 (March 1971), Table C-2.

ers (UAW), which has contracts with all the major farm equipment manufacturers as well as with Caterpillar. It had competition for a number of years with a then rival CIO union, the Farm Equipment Workers (FE) and the FE-UE when FE merged with the United Electrical Workers in 1950. In 1954 and 1955 most of the FE-UE locals in these industries joined the UAW. The other two leading unions in these industries are the International Association of Machinists and the United Steelworkers of America.

INDUSTRIAL LOCATION

When Cyrus McCormick moved his manufacturing operations from western Virginia to Chicago, it was in recognition of where the market lay. Today, the farm equipment industry remains highly concentrated, not only in the Midwest, but in three midwestern states—Illinois, Iowa, and Wisconsin. In 1969, 53.6 percent of all farm equipment employees were found in these states (Table 7). In only one other state, California, is there found any significant number of employees in this industry.

The construction machinery industry is also highly concentrated in the Midwest, although not in so few states. Table 7 shows that about one-third of the industry's employment is found in the same three states where farm machinery manufacturing

TABLE 7. *Farm Equipment and Construction Machinery Industries, Employment by Principal States, 1969*

	Farm Equipment		Construction Machinery	
State	Number of Employees (000)	Percent of Total	Number of Employees (000)	Percent of Total
United States	135.4	100.0	294.1	100.0
Illinois	30.1	22.2	67.2	22.9
Iowa	25.9	19.1	8.8	3.0
Wisconsin	16.6	12.3	19.8	6.7
Total three states	72.6	53.6	95.8	32.6
California	3.2	2.4	14.4	4.9
Michigan	n.a.	n.a.	17.4	5.9
New Jersey	n.a.	n.a.	6.6	2.2
New York	n.a.	n.a.	9.3	3.2
Ohio	n.a.	n.a.	29.6	10.1
Pennsylvania	n.a.	n.a.	20.9	7.1
Total nine states	n.a.	n.a.	194.0	66.0

Source: U.S. Bureau of Labor Statistics, *Employment and Earnings, United States, 1909-70,* Bulletin No. 1312-7 (Washington: Government Printing Office, 1971); and *ibid., Employment and Earnings, States and Areas, 1939-69,* Bulletin No. 1370-7 (Washington: Government Printing Office, 1970).

is concentrated and approximately another 30 percent in the adjacent states from Michigan to New Jersey. Within these states, most of the factories are located along the Great Lakes from Buffalo to Milwaukee and other Wisconsin cities, except for plants in downstate Illinois, in Iowa, and in New Jersey.

Besides these major concentrations, there are a few plants in the South (International Harvester in Memphis and Louisville, LeTourneau in Longview, Texas), and in California (International Harvester, Caterpillar, and Deere). Except in California, these plants involve relatively few employees compared to those found in the Midwest.

Within the Midwest plant locations are found in a few areas where large numbers of Negroes live, such as Cleveland and Chicago, but this is the exception, not the rule. Most facilities are now located in smaller towns, and the trend is definitely in that direction. International Harvester has closed four of its six former Chicago plants as obsolete. Caterpillar, Deere, Allis-

Chalmers, Case, and most other companies are headquartered, and/or concentrated, in cities with relatively small Negro populations: Peoria, Joliet, Decatur, Moline and Rock Island, Illinois; Davenport and Waterloo, Iowa; Milwaukee and Horicon, Wisconsin; and Erie, Pennsylvania. Plant location is certain to be a serious deterrent to large scale employment of Negroes in these industries. As will be pointed out in the chapters which follow, plant location is a significant reason why Negro employment continues at a relatively low percentage compared with such industries as automobiles and steel which are also heavily concentrated in the Midwest,[10] but are in the urban centers with large Negro populations.

[10] See Northrup, *et al., op. cit.,* Parts Two and Four.

CHAPTER III

Negro Employment to 1960

The history of Negro employment in the period prior to 1960 for these two industries is heavily affected by developments at International Harvester, which is not only the largest company in the two industries, but by reason of its early growth and Chicago location, has historically been by far the largest employer of Negroes. We are fortunate also that the records of this company have been preserved; this chapter will rely heavily on the author's previous research,[11] supplemented by materials concerning other companies, and by other published studies.

FROM WORLD WAR I TO 1940

As in many industries, Negroes first came into the farm equipment industry when World War I cut off the flow of immigrants who had heretofore supplied the unskilled labor for America's growing economy. Located in Chicago, a focal spot for the growing northward black migration, International Harvester became a key employer of Negroes.

Historically, Harvester, like many other companies, tended to regard Negroes as just another immigrant group. The most recent immigrants were always considered marginal workers and were given the most unpleasant tasks, but even those only in periods of labor shortages. Thus by the 1870's payroll records of the International Harvester Company reveal that the dominant nationality in the foundry was the Irish. With expanding employment and improved educational opportunities, the Irish gradually migrated out of the foundry. By 1903 the Polish had made heavy infiltrations into the foundries of this company. In 1916 they were still there.

When the World War I labor shortage brought the Negro to Chicago, his first employment at International Harvester, like

[11] The materials pertaining to International Harvester are found in Robert Ozanne, *A Century of Labor-Management Relations, op. cit.*, pp. 183-193.

that of the Irish and Polish, was in the foundry, beginning at the very bottom as helper, chipper and general laborer. By the relatively full employment era of the mid-1920's, the Polish workers were leaving the foundry for cleaner, less arduous jobs. Meanwhile, Harvester was meeting the labor shortage at its Chicago twine mill with large numbers of Negro women.

The Harvester industrial relations department kept mill management posted on the growing percentage of Negroes in its foundries and the superintendents of the various plants discussed the matter periodically. Since past experience with immigrant groups had taught International Harvester that too large a number of any group caused them to club together and form unions, the Harvester superintendents who had the highest percentages of Negroes were ordered to reduce the number, not through dismissals, but through changes in hiring policies. Plants employing no Negroes were not, however, ordered to alter their policies.

In 1925, Cyrus McCormick III requested the company's industrial relations department to make a study of Negro employment in the Chicago plants to determine if a larger number of blacks should be utilized. The resultant report, reproduced below, consisted largely of the opinions of the various plant superintendents, but it was quite favorable to the Negro employees. It concluded:

1. That the works are not receiving applications for common labor jobs from American born white persons.

2. That there are very few if any European immigrants applying for common labor but the Negro is considered equally efficient as the foreign born whites. In some instances the Negro is held to be suitable for semi-skilled work. In one instance a superintendent found that Colored help in his foundry resulted in better costs and better quality than formerly was the case with foreign immigrant employees. The large number of Negro applicants permits of a good selection. Colored employees understand the English language and endeavor to comply with instructions. They are considered more loyal than foreigners. Steel mills are more satisfied with Mexicans for common and semi-skilled labor.

3. No racial trouble between Mexicans, Colored, and white is being experienced.

4. Colored employees do not develop into skilled mechanics. One or two superintendents felt that if employed for a period of years a few might become skilled. The Mexicans at the steel mills are developing into semi-skilled tradesmen but none are employed in mechanical or electrical trades.

5. All superintendents believe that Negroes and Mexicans indulge less in intoxicating liquor than do white employees.

6. The Colored and Mexican employee is less troublesome on the wage and labor question than is the white immigrant.

7. As concerns absenteeism and tardiness the Negro is found no worse a violator than the white employee, American or foreign. Mexicans are seldom absent and are considered more stable than whites.

8. At the Deering Works it has been found that Colored employees suffer fewer injuries than a proportionate number of whites. Similar information was not obtainable from other works.

9. Negroes and Mexicans are no worse trouble makers than other employees.

10. No racial distinction is made in handling Mexican and Colored employees as far as washroom, locker room, job preference, etc. are concerned.

11. The careful selection that the present large number of Colored applicants permits of is resulting in securing an intelligent class, probably of higher intelligence than the white immigrant who does not speak English.

12. When asked whether Negroes or Mexicans were preferred the superintendents were not ready to answer because of limited experience with Mexicans. They liked Negroes, however, because of being able to speak English and being larger in stature. The steel mills have not tried Colored help but are of the opinion that Mexicans will more readily develop into semi-skilled employees.

13. Negroes and Mexicans work in all departments excepting the strictly skilled mechanical trade departments. Generally, however, they fit into semi-skilled or common labor jobs.

14. The superintendents did not desire too many employees of one nationality, preferring a mixture of nationalities unless American employees are largely American born.[12]

This study was presented to the works managers in 1925, but despite its favorable conclusions in regard to Negro employment, it did not seem to have had any policy impact. Cyrus McCormick III did not follow up his interest, and policy was left largely to the superintendents subject only to an occasional directive that the proportion of Negroes in some plants was becoming too large. In 1928, of the Chicago plants, the Webber Works and McCormick Twine exceeded a 20 percent black ratio and the McCormick

[12] Ozanne, *op. cit.*, pp. 186-187.

Works was close to that figure, with Deering and the Tractor Works both about 10 percent Negro. But the Wisconsin Steel Works, which employed a large percentage of Mexican-Americans, the Deering Rolling Mill, and the West Pullman Works employed absolutely no Negroes because of the wishes of their superintendents, their location far to the South of the then Negro housing district, and the possibly more virulent prejudices of far South Chicago residents. This hands-off policy by central administration continued until World War II when a new critical labor shortage forced these lily-white plants to employ Negroes. In the meantime, the layoffs occasioned by the Great Depression hit Negroes disproportionately hard. Thus in the McCormick Twine Mill, where Negro employment reached a peak of 27.5 percent in 1929 and in the McCormick Works where it stood at about 18 percent in 1923, the percentages had dropped by 1940 to 18.8 and 10 percent, respectively.

Other manufacturing concerns had similar experiences. In the quad-city agricultural implement center lying on both sides of the Mississippi River [13] and extending out to the prairie cities in both Iowa and Illinois, the World War I labor shortage brought Negroes to the foundries of Deere and Company. The company built houses to handle this influx.

As far north as Milwaukee, Allis-Chalmers could man its foundry in World War I only by hiring Negroes. The hiring dates of the 101 Negroes employed in the Allis-Chalmers West Allis Milwaukee foundry at the beginning of 1941 indicate that although a Negro furnace tender had begun work there in 1898, this company seemed to have added Negroes only as a last resort in periods of labor shortage such as during World War I and the early 1920's.[14] The fact that they were all in the foundry prior to 1942 not only indicates that they were employed in the least desirable jobs, but implies a practice of restricting them to the foundry.

During this early period, construction machinery was still an infant industry for which separate census employment data were not maintained. We do, however, have separate census data for the farm equipment industry, which are set forth in Table 8.

[13] In these cities of Moline and Rock Island, Illinois and Davenport and Bettendorf, Iowa are a number of farm equipment plants, as well as the headquarters of Deere.

[14] Data from the company.

TABLE 8. *Farm Equipment Industry*
Total Employed Persons by Race and Sex
United States, 1910-1960

Year	All Employees			Male			Female		
	Total	Negro[a]	Percent Negro	Total	Negro[a]	Percent Negro	Total	Negro[a]	Percent Negro
1910	37,620	596	1.6	36,012	594	1.6	1,608	2	0.1
1930	53,244	1,092	2.1	50,140	1,080	2.2	3,104	12	0.4
1940	91,140	1,420	1.6	85,760	1,420	1.7	5,380	—	—
1950	179,250	7,290	4.1	159,690	7,080	4.4	19,560	210	1.1
1960	130,696	5,497	4.2	117,059	5,297	4.5	13,637	200	1.5

Sources: *U. S. Census of Population:*
1910: Vol. IV, *Occupation Statistics*, Table VI.
1930: Vol. V, *Occupations-General Report*, Chapter VII, Table 1.
1940: Vol. III, *The Labor Force, Industrial Characteristics*, Table 1.
1950: Special Report P-E No. ID, *Industrial Characteristics*, Table 2.
1960: PC (2) 7F, *Industrial Characteristics*, Table 3.

Note: Workers 10 years old and older in 1910 and 1930, 14 years old and older, 1940-1960.

[a] Figures are for nonwhite in 1940.

Although Negro employment almost tripled between 1910 and 1940, it merely kept pace with total industry employment, losing the gains made prior to 1930 during the depression decade. Undoubtedly the bulk of the Negroes reported by the census to be employed in the farm equipment industry during the 1910-1940 periods were employed in the Chicago plants of the International Harvester Company.

THE WORLD WAR II PERIOD

International Harvester, like many companies during World War II, opened its doors to Negroes beyond any previous extent. The shortage of labor during the war period was one reason for the change; another was the adoption by the federal government of a nondiscriminatory employment policy. Executive Order 8802 of June 1941, issued by President Franklin D. Roosevelt, established a committee on Fair Employment Practice (FEPC) to enforce the requirement—the first in a long line of executive orders in this regard. An immediate result was the company decision to centralize its racial employment policies. Table 9 shows the results: not only a sharp increase in the proportion of Negroes hired in the plants where their employment had a long history, but a dramatic termination of racial exclusion at the Wisconsin Steel and West Pullman Works. With this change began a program of nondiscrimination, which until the 1960's, was probably unrivaled in American industry. It was basically attributable to two people:

> One was Sara Southall, a member of the Harvester industrial relations staff. For some years after she joined Harvester she worked as a volunteer with the industrial department of the YWCA and with the Chicago Urban League, where her efforts were directed toward fair employment. Because of her knowledge and experience in this field she was asked to serve on President Roosevelt's Fair Employment Practices Committee. The Harvester Company granted her leave for this assignment. Awareness of Miss Southall's work on the FEPC inevitably tended to focus more interest of Harvester executives on the company's own performance in this area. However, no compelling, companywide program for fair employment would have been possible without the interest and vigorous leadership of Fowler McCormick, who became president of International Harvester in 1941. McCormick held a deep conviction that people should be dealt with as individuals, not as groups. He vitalized this belief by insisting on a fair employment policy in all Harvester plants throughout the country.[15]

---

[15] Ozanne, *op. cit.*, p. 188.

TABLE 9. *International Harvester Company*
Percent Negro Employment by Plant
1940-1960

Year	West Pullman	Wisconsin Steel	McCormick	Tractor	McCormick Twine	Melrose Park	Louisville	Memphis	Company Total
1940	*	*	10.3	6.5	18.8	†	†	†	4.5
1941	*	*	*	*	*	†	†	†	*
1942	0.5	11.2	11.2	7.3	29.9	†	†	†	5.9
1943	9.7	13.4	19.3	12.9	51.9	†	†	†	9.5
1944	10.2	11.1	24.8	18.9	62.1	†	†	†	11.6
1945	7.4	13.3	25.3	17.5	63.6	†	†	†	11.7
1946	6.4	10.0	26.0	18.5	67.0	4.0	4.2	†	11.0
1947	8.4	13.2	30.3	18.6	70.4	8.5	6.5	12.2	11.5
1948	8.6	17.2	31.5	19.4	59.6	10.1	5.9	19.6	11.9
1949	7.9	14.4	19.3	19.3	52.7	10.5	9.5	23.2	10.6
1950	8.9	17.0	28.5	22.4	64.8	12.8	14.1	23.1	12.8
1951	9.8	18.6	20.5	23.6	75.6	13.5	14.3	21.1	12.7
1952	12.7	19.7	31.4	24.1	42.4	14.1	16.9	20.5	13.8
1953	5.5	14.4	13.6	16.2	†	11.8	11.2	26.1	9.8
1954	10.7	17.3	24.1	17.5	†	14.4	13.2	24.8	11.3
1955	12.0	19.1	33.0	19.3	†	16.6	20.9	23.0	14.5
1956	10.7	18.3	25.7	22.1	†	18.1	20.1	23.8	14.1
1957	12.4	17.4	32.6	18.5	†	14.5	16.7	23.3	13.0
1958	11.5	18.6	34.5	17.9	†	16.0	15.4	21.6	13.5
1959	12.3	16.9	37.1	25.8	†	18.4	14.4	23.5	13.8
1960	9.4	13.4	28.7	13.3	†	9.3	11.9	19.3	9.3

From Robert Ozanne, *A Century of Labor Management Relations at McCormick and International Harvester* (Madison: University of Wisconsin Press; © 1967 by the Regents of the University of Wisconsin), p. 192.

* Figures not available.

† Plant not yet built, or closed.

During World War II, Allis-Chalmers consciously began a gradual process of introducing Negroes to all sections of its Milwaukee area operations. Management moved with caution since it anticipated opposition from its white workers. Many superintendents and department heads opposed Negro employment, warning of possible trouble with the skilled trade unions and white workers. Consequently, the first Negroes hired were exceptionally well qualified. The results were not spectacular. In 1942 the first female Negro clerical was hired into a plant office and, for the sake of high visibility, was put in the industrial relations section. Not until the early 1950's were Negro women employed in the Allis-Chalmers main office. Although Negroes were introduced into almost all the departments of the plant, they were not placed in top jobs nor as foremen. As late as 1946, out of a total of 41 Negro hires in the month of April, 38 were put in the foundry. In May of 1948, 68 percent of the foundry Negroes were in the laborer classification, whereas in the high paying jobs of molder and coremaker, 74 percent and 97 percent, respectively, were white.[16] Table 10 shows the percentage of Negroes in the Milwaukee area operations of Allis-Chalmers, 1942-1960.

During World War II, Negroes secured their first jobs in such companies as J. I. Case and Massey-Harris, and in other manufacturing concerns in the Mississippi Valley cities where farm equipment concerns are concentrated. Most of the black employees hired at this time were used in the foundries and remained there. At the Deere and Company plants, which had first employed Negroes during World War I, the number and proportion of black employees increased, but most remained in unskilled and/or foundry jobs.

At Caterpillar Tractor, headquartered in the central Illinois city of Peoria, which is eighty miles east of the Mississippi River and off the regular migration routes from South to North, special interest in Negro employment began in 1941, prodded by the Interracial Fellowship of Peoria. This group inquired why so few Negroes were on the company's payroll. The company agreed to employ any qualified persons referred by the group, and did so.[17] During World War II also, the construction

[16] Data from company.

[17] Stephen Habbe, *Company Experience with Negro Employment,* Personnel Policy Study No. 201 (New York: National Industrial Conference Board, 1966), Vol. I, p. 104.

TABLE 10. *Allis-Chalmers Manufacturing Company Percent Negro Employment West Allis Plant and Main Office, 1942-1960*

Year	Percent Negro Employment
1942	0.7
1943	1.1
1944	1.6
1945	2.0
1946	1.6
1947	n.a.
1948	n.a.
1949	3.5
1950	3.5
1951	4.4
1952	5.4
1953	7.0
1954	5.2
1955	5.9
1956	5.7
1957	5.2
1958	5.3
1959	6.2
1960	5.4

Source: Company data.

machinery industry was given a tremendous boost by orders for equipment for war use. The tight labor market, the government's new insistence on fair employment, and the migration of Negroes from southern agriculture to northern industry resulted in employment of a substantial number of blacks in plants of this industry especially in the cities along the Great Lakes from Chicago to Erie. Most, of course, were used in unskilled and foundry work, but as in the automobile industry, others obtained jobs on the assembly lines and as machine tenders in parts manufacturing departments.

FROM WORLD WAR II TO 1960

The period between World War II and 1960 saw Negroes hold on to some of their wartime gains in farm equipment and construction machinery manufacturing, increase them in some facilities, but lose out in others. International Harvester now moved to increase black employment.

"Under Miss Southall's planning and guidance and Fowler McCormick's personal attention the company quietly developed a comprehensive plan for equal employment opportunity at the end of World War II." [18] The data in Table 9 show that in the immediate postwar years the black workers' share of employment at Harvester rose from 11.6 percent in 1945 to 13.8 percent in 1952. In the following year, the McCormick Twine plant in Chicago ceased production. There Negro employment had reached a peak of 75.6 percent in 1951, and stood at 42.4 after initial closing layoffs in 1952. The demise of this plant reduced the overall Negro ratio in Harvester to 9.8 percent, but it quickly moved upward again.

Harvester in the South

A major reason why the black overall percentage at Harvester expanded after the closing of the McCormick Twine plant in 1952 is attributable to the courageous and innovative policy pursued by the company at its then newly opened southern and border city locations. Following World War II, Harvester acquired major surplus facilities in Evansville, Indiana [19] and Louisville, Kentucky, and built a new plant in Memphis, Tennessee. Although segregation was common in the first two cities and almost ironclad in the third, Harvester employed, trained, and upgraded on a nondiscriminatory basis. This did not result in equal employment, but it did move closer to that ideal in the deep South city of Memphis probably more than in any other major manufacturing facility at this time.

President Fowler McCormick set the management tone for the southern venture at a Chamber of Commerce banquet celebrating the opening of the Memphis plant, where he told the assembled political, civil, and business leaders, "It is our belief, in our company, that all of us have a responsibility to hire colored people, . . . We must give them an opportunity to earn a living. It is our company policy." [20] He further emphasized that Negroes

[18] Ozanne, *op cit.*, p. 188.

[19] Harvester used the Evansville plant as the basis for its post World War II entry into the household refrigerator-freezer business. In the mid-1950's, it ceased operations in this field and sold the plant to the Whirlpool Corporation.

[20] Ozanne, *op cit.*, p. 189.

would be employed at "the same wages for the same work" as would whites.[21]

The company was careful to advise new employees of its policies. When the Memphis plant was established, white applicants were told of the company's policy of integration and were urged not to accept jobs in the plant if they were strongly opposed to it. This integration policy has been continued since, although with somewhat varying degrees of emphasis.

To promote the success of its program, Harvester became one of the first major concerns to centralize its racial employment policies. The racial employment status of each plant and the progress for each period were submitted by plant managers to the company executive committee. Few major concerns at this time regarded racial employment policy to be a top management concern requiring close and effective top management supervision.

The United Auto Workers became the bargaining representative and from its Detroit and Chicago offices its international representatives and officers expressed consistent support of nondiscrimination clauses in the contract and of enforcement of equal job opportunity regardless of race, as will be pointed out below. Since Negroes were hired from the inception of the plant, the white employees had no seniority advantage over Negroes, unlike the situation in so many older plants where the Negroes were latecomers.

These policies conducive to employment integration were, of course, pitted against the traditional segregation patterns of a city located in the deep South. At the beginning these traditions included segregated plant toilet and cafeteria facilities, segregated community education, and strong local support among the white workers, foremen, local managers, and local union officers for this segregation.

International Harvester has given access and assistance to several independent scholars to make studies of the program of employment integration at the Memphis plant. The most comprehensive study of racial employment practices at the Memphis plant was made by John Hope II, who also analyzed the situation at Louisville and Evansville. His study covers the period from World War II to the early 1950's.[22]

---

[21] *Ibid.*

[22] John Hope II, "Negro Employment in 3 Southern Plants of International Harvester Company," in *Selected Studies of Negro Employment in the South,* Report No. 6, NPA Committee of the South (Washington: National Planning Association, 1953).

Mr. Hope's data are summarized in Figure 1 which shows the percentage of Negro employees in these plants and in Figure 2 which shows the distribution of these Negro employees in the various skill levels. Negro employment increased in the company generally during this period, up to 23 percent in the Memphis plant. Mr. Hope reported that, "In compliance with the nondiscrimination policy, hundreds of Negroes have been employed in a variety of semiskilled production jobs and have received promotions to higher job classifications on the basis of their qualifications and seniority without regard to their race."[23] Speaking more specifically Mr. Hope stated:

> By the end of 1950, about two-thirds of the Negro employees at both the Memphis and Evansville Works and a little more than two-fifths of those at the Louisville Works were engaged in jobs classified above the unskilled level traditionally designated for Negroes in these communities. All of these were semiskilled workers at Memphis, while there were a few Negroes in skilled classifications at Louisville, and in the clerical, professional, technical, and managerial category at Louisville and Evansville. Negroes in this latter category at Evansville in 1950 included three clerks, one chemist, one draftsman, and one typist. In previous years Negroes had also held the position of messenger, laboratory assistant, and employment counselor. At Louisville this category is represented by one Negro employment interviewer.[24]

Mr. Hope noted, however, that despite these gains the overwhelming majority of Negroes were employed in the foundry. With one exception these three plants "have not employed Negro journeymen or journeymen's helpers in the apprenticeable trades, nor have they enrolled Negroes in Company sponsored apprenticeship training." [25] In fact, Mr. Hope's report demonstrates that, in violation of the union contract and company policy, local union and company officials, under pressure from white workers, kept Negro union members out of machine jobs, welding, inspection, and supervision over whites.[26] An example from Mr. Hope's study demonstrates this local sabotage.[27]

In March 1948 before the UAW had bargaining rights, a Negro applied for a job as a welder at the Memphis plant, passed the required test, and was told to report to work. Upon hearing that a Negro was taking a test for welding, the white welders pro-

[23] *Ibid.*, p. 126.

[24] *Ibid.*, p. 37.

[25] *Ibid.*, p. 126.

[26] *Ibid.*, pp. 101, 107, 110-111.

[27] *Ibid.*, pp. 105 and 106.

FIGURE 1. *International Harvester Company*
Percent Distribution of Negro Employment
Selected Locations and Company Divisions, 1946-1950

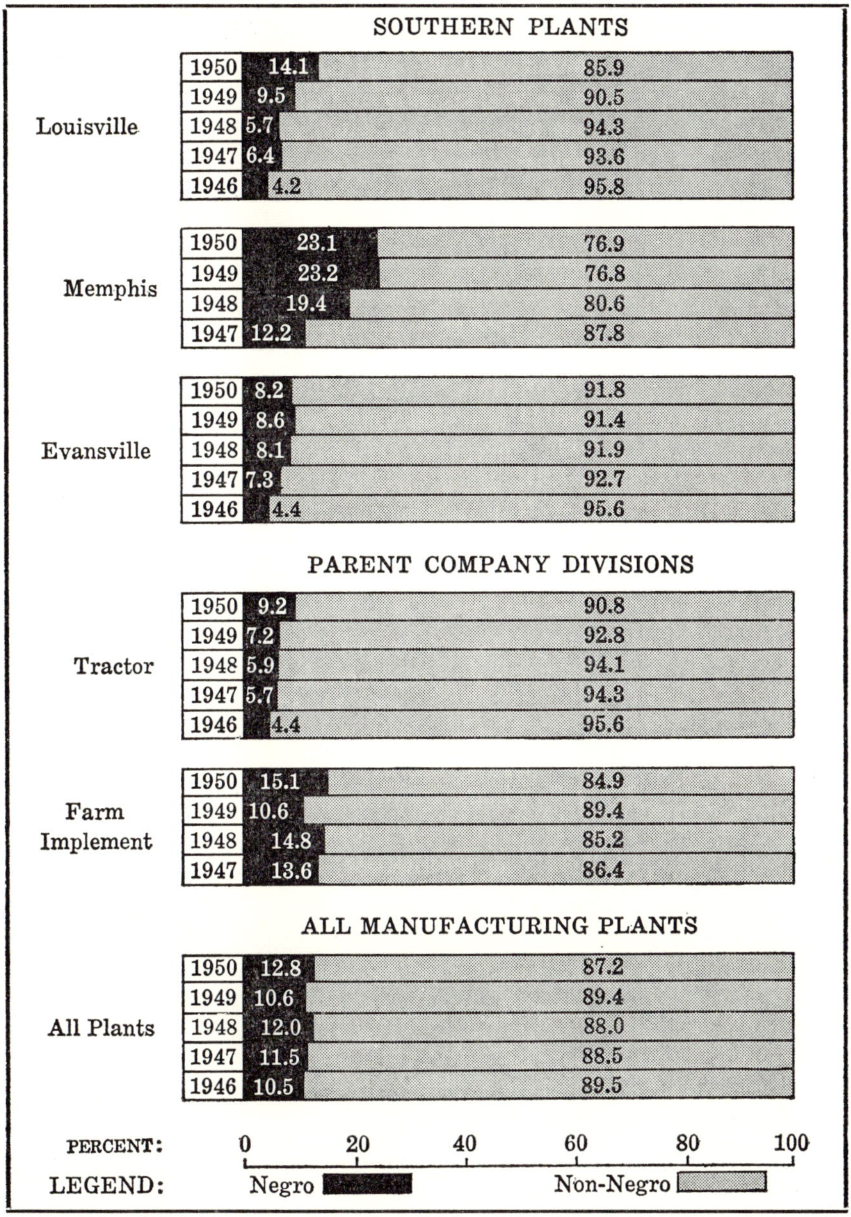

Reprinted from John Hope II, "Three Southern Plants of International Harvester Company," in *Selected Studies of Negro Employment in the South,* NPA Committee of the South, Report No. 6 (Washington: National Planning Association, 1955), p. 35, by permission of the publisher. © 1953 by the National Planning Association.

FIGURE 2. *International Harvester Company*
Percent Negro Employment by Skill Levels
Selected Locations and Company Divisions, 1946-1950

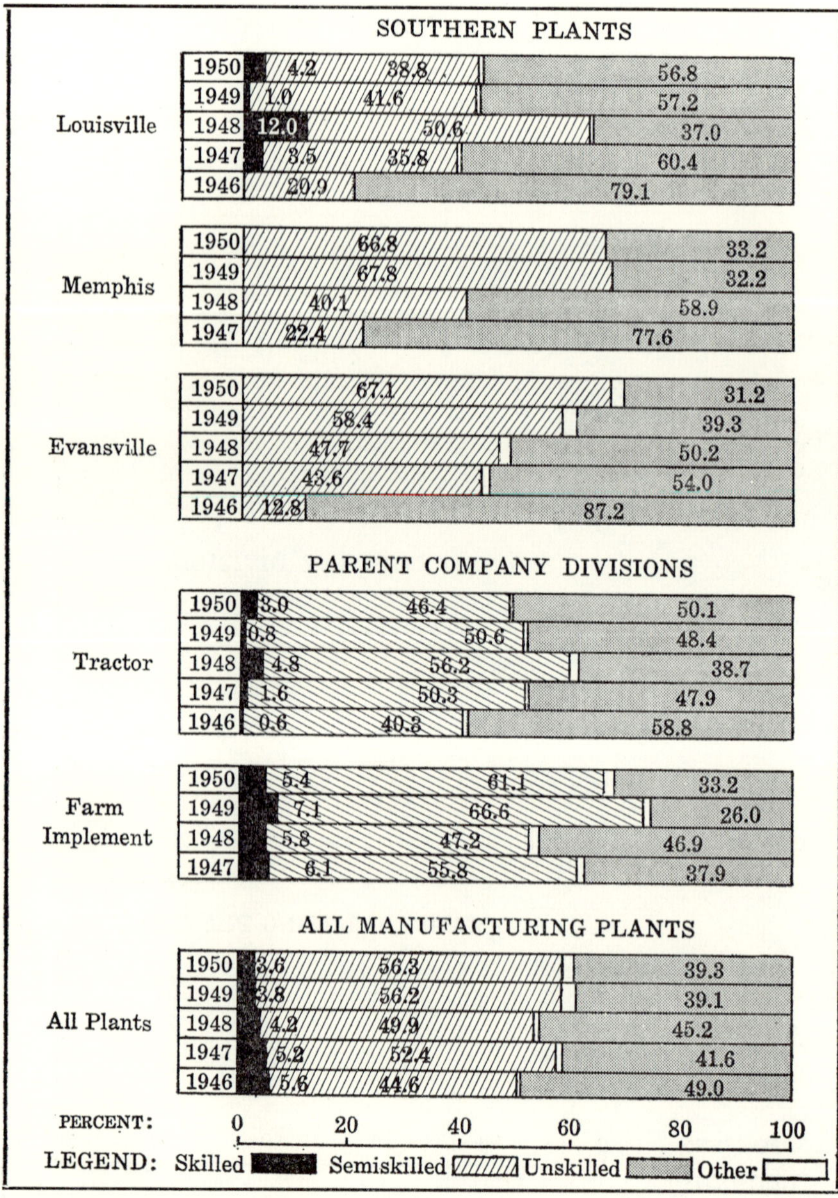

Reprinted from John Hope II, "Three Southern Plants of International Harvester Company," in *Selected Studies of Negro Employment in the South,* NPA Committee of the South, Report No. 6 (Washington: National Planning Association, 1955), p. 38, reprinted by permission. © 1953 by the National Planning Association.

tested with an hour-long sit-down. When the Negro reported for work as a welder he was asked to take temporarily a laborer job because no booth was ready for him. In spite of his frequent inquiries no booth was ever ready for him and he worked as a laborer for the ensuing year. When the UAW gained bargaining rights in April 1949 he presented his grievance to the new local. Now the union delayed, "lost" his written grievance, and finally, at the insistence of other Negro union members, held a meeting with the UAW Fair Employment Practice Department representative, the local union officers, and the grieved Negro. The Negro was persuaded to forego his claim as a welder to avoid the destruction of the new union.

Two years later after two new white welders had been hired, this Negro again filed a grievance. This time the company agreed to "give him his booth," but the union still feared violence. The union president refused to hold a joint Negro-white union meeting on the grievance but met separately with some Negro members. They agreed to wait another 30 days during which time the union president would educate the white welders to the rights of the incoming Negro welder. There are no further details on this case except that two years later this same Negro was working as a core-maker in the foundry but with the same grade and wage rate as a welder. He had been denied the personal satisfaction and status of being a welder.

Harvester's attempts to adhere to its policy of equal employment opportunity in the South in the post-World War II period were seriously hindered by the absence of training opportunities for Negroes in the segregated schools. Mr. Hope reported:

> A comparison of vocational courses offered to White and colored reveals a conspicuous absence in Negro schools of mechanical and metal-working courses—of machine-shop, sheet metal, steamfitting, plumbing, and other courses of value in the modern manufacturing plant. The Harvester plants in the cities studied are virtually the only places where Negroes can obtain even the basic machine-shop training necessary for semi-skilled jobs.
>
> On the other hand, White public school facilities of various sorts furnish courses of training ranging from those providing a general acquaintance with the shop environment, tools, and practices to vocational courses which aim at providing the student with a skill or trade with which he may make a living. With such instruction the student can qualify for apprenticeship training courses, upon completion of which he can gain journeyman status in highly skilled craft unions.[28]

[28] *Ibid.*, p. 28.

A further summary of the vocational training situation in the three Harvester plant cities is provided in Table 11. Obviously companies desiring to employ on an equal basis found it difficult to overcome years of neglect and discrimination, factors still limiting Negro employment in the skilled trades. Nevertheless, Mr. Hope ended his study optimistically believing the progress would continue.

A second detailed study of Harvester's Memphis plant was made by Professor Robert Weintraub, who analyzed the wage differences between white and Negros employed there to determine whether the nondiscrimination policy in employment had been successfully enforced and therefore whether efforts of the management to integrate Negroes could be effective in a southern community.[29] The company provided elaborate wage and personnel data on a substantial sample of workers covering their employment from September 1, 1947 through March 1, 1953.

Mr. Weintraub's conclusions follow:

1. The effort was at least partially successful since the wage difference between Negro and white metal workers at the Memphis plant was not much larger than that in the country as a whole and clearly smaller than the disparity in earnings between the two races in other southern metalworking plants.
2. The wage difference which persisted in the Memphis plant could be partially accounted for by the difference between whites and Negroes in the amount of previous job experience or vocational training.
3. The remaining wage differential was probably due to discrimination (or as he used the term "resistance to integration"), since during the period studied not a single Negro in the sample was assigned to a supervisory position or to a skilled job.

Like Mr. Hope, Mr. Weintraub ends on the optimistic note, "that the terminating date of this research was March 1, 1953 and the company's nondiscrimination policy is a continuing effort which probably has achieved additional successes in carrying integration into the Memphis plant." [30]

---

[29] Robert Weintraub, "Employment Integration and Racial Wage Differences in a Southern Plant," *Industrial and Labor Relations Review,* Vol. 12 (January 1959), p. 214-226.

[30] *Ibid.,* p. 225. Of interest also is Professor Weintraub's other article "The Productive Capacity of Rural and Urban Labor," *Journal of Political Economy,* Vol. LXIII (October 1955), pp. 412-426.

TABLE 11. *Work Related Vocational Training in Public Schools Three Cities, 1950-1951 School Year*

	Day School			Adult Evening School					
	Vocational Traning Courses			Apprenticeship Training Related Courses			Vocational Trade Extension Courses		
	Louisville	Memphis	Evansville	Louisville	Memphis	Evansville	Louisville	Memphis	Evansville
Number of classes									
White	17	12	8	9	22	26	12	12	48
Negro	6	7	4	0	3	0	6	3	1
Number of subjects									
White	13	9	6	9	10	9	7	12	18
Negro	4	6	4	0	3	0	4	2	1

Reprinted from John Hope II, "Negro Employment in 3 Southern Plants of International Harvester Company," in *Selected Studies of Negro Employment in the South*, NPA Committee of the South, Report No. 6 (Washington: National Planning Association, 1955), p. 30, by permission of the publisher. © 1953 by the National Planning Association.

The present author's research included a study of the Memphis and Louisville plants; the percentages of Negroes employed in these two locations are shown in Table 9. The fluctuations in Negro employment at Memphis after 1950 (a high of 26.1 percent in 1953 to a low of 19.3 percent in 1960) as well as in the entire company (a high of 14.5 percent in 1955 to a low of 9.3 percent in 1960), indicate the difficulties besetting black employees even when a company consciously attempts integration.

Although International Harvester substantially increased its proportion of black employees in the forties and early fifties, it did not do so proportionally in all departments. Notably, the black employment in the skilled metalworking departments and the crafts lagged behind that in the assembly and foundry departments. When layoffs came in 1960 there were more in the assembly and foundry than in the tool room and maintenance. Thus, black layoffs were proportionately greater than white. This occurred at Memphis and Louisville even though the blacks had equal seniority with the whites. In certain older Harvester plants which had been lily-white before World War II the disproportionate decline of black employment was caused by the lesser seniority of blacks.

The author summarized the achievement of Harvester's racial policies at the end of the 1950's.

> Perfect integration has not been achieved, but the company has moved forward, reaching a companywide percentage of Negro employment of 13.8 in 1959 compared to 4.5 in 1940. . . . In most plants the ratio of Negro to white employees is about equal to the ratio in the community. There is still a higher level of Negro employment in plants that have foundries, and a higher percentage of Negroes in unskilled work; but large numbers of Negroes are employed in semiskilled work. In 1959, 207 Negroes were in skilled apprenticeable occupations throughout the Harvester plants. Gradually a few Negroes have been put in salaried jobs and clerical jobs, even in the Southern plants. In the general office in Chicago in 1959 thirteen Negroes were in positions of clerk or above. Locker rooms, toilets, and eating facilities have all been integrated.[31]

The implementation of a national equal employment opportunity program, applied to southern as well as northern plants, with affirmative action aspects under centralized direction, was Harvester's unique contribution in the years before 1960.

An interesting aspect of the Harvester equal employment opportunity program is that it was one of three notable pioneering programs during the years before a national policy developed

[31] Ozanne, *op. cit.*, p. 193.

which was adopted by major companies under the leadership of a founding family. The other two, those at Ford Motor Company and Firestone Tire and Rubber Company, were instituted by the founders. The great pioneer aspects of Harvester's policy were instituted by Fowler McCormick, a direct descendant of the founder.

Harvester was not the first company to go beyond equal employment by insisting on the affirmative action of recruiting, training, and upgrading Negroes, nor was it the first to centralize racial employment policies. Ford, under its founder's aegis, had done much the same in the 1920's and early 1930's. Ford, however, confined that policy to its Highland Park and giant River Rouge Works, both in the Detroit area. Its southern assembly plants made no effort before 1960 to contravene the mores of the South. In these facilities Negroes prior to that date were confined to menial jobs.[32]

"Like his friend Henry Ford, Harvey Firestone, Sr. founder of the company that bears his name, became deeply committed to giving Negroes a share of employment. Firestone not only was the first [in the rubber tire industry] to employ Negroes, but throughout most of the industry's history, has consistently had the highest percentage of Negroes of any Akron plant, and in its Memphis, Tennessee plant, the highest percentage in any plant in the country." [33] In Memphis, however, Firestone employees were confined to segregated departments until 1957, at which time a plant seniority program was implemented allowing Negroes to bid on jobs throughout the plant.[34] This was long after the Harvester plant in Memphis was well integrated. The implementation of a national equal employment opportunity program, applied to southern as well as northern plants, with affirmative action aspects under centralized direction, was Harvester's unique contribution in the years before 1960.

The General Picture in the 1950's

The proportion of Negroes in the farm equipment industry tended to remain virtually stationary at a little more than 4 percent during the 1950's. (See Table 8.) This occurred despite improvement at Harvester and at the Milwaukee facilities of Allis-

---

[32] Northrup, *et al., op. cit.,* Part Two, pp. 55-58.

[33] *Ibid.,* Part Five, pp. 401-402.

[34] *Ibid.,* pp. 422-423.

Chalmers (Tables 9 and 10). The latter company continued its wartime equal employment program in the postwar years, substantially increasing the percentage of Negroes in its work force, slowly moving them into all plant departments, and employing them in office positions as well. At this time, however, Allis-Chalmers did not commit itself to centralized direction of affirmative action, as its experience with federal and state government investigators during this period showed.

The first effective peacetime federal efforts to encourage jobs for Negroes were begun in the 1950's. Although the Eisenhower Administration did not support federal fair employment legislation, a President's Committee on Government Contracts was established on August 13, 1953 by Executive Order 10479 to encourage nondiscrimination in employment by government contractors. Known as the "Nixon Committee," because it was chaired by the then Vice-President, this committee developed investigatory and fact-finding techniques, later expanded by the Kennedy and Johnson Administrations, as a means of increasing Negro employees among government contractors. Plants of Allis-Chalmers in Norwood, Ohio, Springfield, Illinois, and Pittsburgh, Pennsylvania were surveyed by Nixon Committee investigators in 1957 and 1959. The files of the Committee noted that "no information of a satisfactory kind could be obtained" for the Norwood and Springfield plants in 1957. Its files show that in 1959 the Pittsburgh plant "employed 275 professional-technical workers, of whom one was a Negro; 253 supervisory employees, of whom none was a Negro; 226 clerical-stenographer personnel, of whom none was a Negro; 237 skilled workers, 766 semi-skilled workers and 450 unskilled workers. The surveying official estimated that there were 74 Negroes distributed among the skilled, semi-skilled and unskilled categories." [35]

Besides the federal activity in the equal employment area, the states had become increasingly a factor since World War II. The early state laws, however, were largely found in states in which few farm equipment or construction machinery plants are found. In the late 1950's, however, antidiscrimination laws were adopted in a number of midwestern states, including Ohio and Illinois. A case in 1959 involved the Ohio agency and the Norwood plant of Allis-Chalmers, from which, as noted above, the Nixon Committee had been unable to obtain data.

---

[35] From unpublished reports of President's Committee on Government Contracts, p. 488.

In the Norwood plant, located in a border city, the company was employing only 94 Negroes out of a total labor force of over 2,000. Nearly all of the 94 were in a labor pool; the only two Negro women were maids; and there was apparently a practice of confining Negroes largely to unskilled work.[36] This situation, of course, differed substantially from that in the same company's plants in Milwaukee and Pittsburgh, and was typical of the variations in company practice which resulted from lack of central control of racial policies during this period.[37]

Caterpillar Tractor's Joliet, Illinois, plant was the only other major facility for which the records of the Nixon Committee indicate that an employment survey was made. In the late 1950's, probably less than 2 percent of the company's total employees were black,[38] but the Joliet plant had a far better record, being located much closer to centers of black population than were nearly all other of the company's plants. Most Negroes were concentrated in semiskilled and unskilled jobs, and none were in supervisory or professional-technical positions.

Earlier we noted the difficulties which Harvester had in recruiting Negroes for skilled jobs in southern and border cities. Caterpillar's experience in Joliet shows that this was not just a southern phenomenon. Thus the investigator reported to the Nixon Committee:

> It is significant that no Negroes are included among 88 apprentices and trainees, most of whom were recruited from Joliet high school where Negro students have not been enrolled in shop training. In this connection, you will recall that we transmitted to you a copy of a special report on March 20, 1957, relating to building trade practices in this area and that this report disclosed that many of the local trade unions do not accept Negro members. Consequently, school counsellors discourage Negro students from entering shop training courses because their acquired skills cannot be properly utilized.[39]

The report noted that the Joliet Mayor's Human Relations Committee, of which the Caterpillar plant labor relations director was

---

[36] From the unpublished files of the Ohio Civil Rights Commission.

[37] For an analysis of similar situations in other industries, see Northrup, *et al., op. cit.,* Part Two, pp. 53-55, 62-63; and Purcell and Mulvey, *The Negro in the Electrical Manufacturing Industry, op. cit.,* pp. 40-42.

[38] Habbe, *op. cit.,* p. 104, reports that Caterpillar had only 637 minority employees as late as February 1963.

[39] From the unpublished files of the President's Committee on Government Contracts, pp. 492-493.

a member, had met with school officials in an attempt to break this discriminatory circle.[40]

In other plants in farm equipment and construction machinery, Negro employment continued to depend heavily on plant location and the state of employment. As the decade of the 1950's ended, neither industry indicated an overly optimistic outlook for black employment. The recessions of 1958 and 1960 caused severe employment declines in both industries. A number of farm equipment plants closed and employment in that industry was substantially less in 1960 than ten years earlier. On the other hand, the construction machinery industry was poised for a great decade of expansion—but one that in 1960 had yet to begin.

UNION RACIAL POLICIES TO 1960

In most manufacturing industries hiring and training have been functions exclusively reserved to management. Historically, even where such management policies were racially discriminatory, unions seldom attempted to interfere because they had no contractual right to do so, because racial prejudices are no respecter of rank, and because white workers generally benefitted economically from racially discriminatory hiring policies. Today, as racial bars are being removed, unions can pose problems for minority employment progress by institutionalizing the status quo; or they can bolster integration policies of management by support. In this series, we have examined notable cases of all such policies.[41] There are also interesting examples in the industries examined herein.

Undoubtedly, the United Automobile Workers has as good a record as any American union in racial matters. Previous studies have discussed its policies in the automobile industry where it has been a strong force for equality [42] and in the aerospace industry

[40] *Ibid.*, p. 493.

[41] See, for example, Northrup, *et al., op. cit.*, Part Two, pp. 99-104; Part Three, pp. 214-216; and Part Five, pp. 481-485; and Herbert R. Northrup, Richard L. Rowan, *et al., Negro Employment in Southern Industry,* Studies of Negro Employment, Vol. IV (Philadelphia: Industrial Research Unit, Wharton School of Finance and Commerce, University of Pennsylvania, 1971), Part One, pp. 37-45, 112-117; Part Three, pp. 17, 33-39, 84-87; and Part Four, pp. 21-23, 34-36, and 59.

[42] Northrup, *et al., op. cit.*, Part Two, pp. 99-104.

where its influence has been far less significant.[43] The present writer described the UAW's role, and particularly the internal union struggle over integration in the Memphis local, in his previous work.

Although the international officers of the UAW gave Harvester unequivocal support for the integration policies of the company, neither membership in UAW Local 988 in Memphis, Tennessee, nor employment in Harvester's integrated plant magically transformed all white workers raised in Tennessee, Arkansas, and Mississippi into crusading integrationists. On the other hand, about 500 of the 1,800 local members were Negroes. Thus a local union meeting was a concentration of the full range of racial feelings.

> As required by the national union, the union meetings were integrated, though the Negro members tended to sit together. The UAW policy of integration was being challenged repeatedly at union meetings by segregationists under the leadership of a union steward who was formerly a vice president of Local 988. The steward's motions were continually ruled out of order by the local president. On an appeal of these rulings Herschel Davis, administrative assistant to UAW Vice President Pat Greathouse, came from Detroit to Memphis. At a union meeting he warned that Local 988 might lose its UAW membership by segregationist activity. He was jeered with shouts of "Go back up North." The leader of the segregationists told Davis: "We belong to Reuther's union but we don't believe in Reuther's integration. . . . We may have to choose between CIO and segregation. For my part, I'll take segregation. The time will come when workers of the South will decide whether we want your union with desegregation or our own union with segregation. When the showdown comes, we'll take segregation and leave you guys up in Detroit."[44]

In 1956 Local 988 constructed a new union hall across the street from the plant. In defiance of orders from Detroit

> . . . signs "White" and "Colored" appeared over restrooms and drinking fountains. Thus began a two-year battle between Local 988 and the national union. After trying to persuade the local to remove the signs, UAW Vice President Greathouse went to Memphis and ordered the signs painted out. When he left Memphis the local ordered new signs painted on. A labor columnist . . . told the story of Pat Greathouse's return to Memphis: "It was late this last Monday night when a car, headlights out, glided up to the curb in front of Local 988 UAW in Memphis, Tenn. Three men, one carrying a brush and a can of paint, climbed out and scurried across the sidewalk to the main entrance of

[43] *Ibid.*, Part Three, pp. 214-216.

[44] Ozanne, *op. cit.*, pp. 190-191.

the darkened building." They painted out the signs, changed the building's locks, and were in physical control of Local 988 when the local's officers returned the next morning.

In describing a subsequent meeting with the local membership Douglas Fraser, member of the UAW Executive Board, said "I have never been called such filthy names in my life. They said they were going to throw both Pat and I into the Mississippi and for a time we thought they were." [45]

Local 988 was placed under trusteeship by the UAW International and the segregationist leadership was replaced by integrationist officials appointed by the UAW International Executive Board. In the bitter intraunion struggle about two hundred segregationist-minded workers, mostly in skilled trades and welding, quit paying dues to Local 988. This attrition enabled the moderates to win elections following the end of the trusteeship in the early 1960's.

In the day to day union-management relations, the pressures on the union and the management led to compromises with the equalitarian programs of both. Hope tells of a number of incidents during the early years of the plant where Negroes were discouraged from asserting their upgrading rights under the contract because of threatened or actual adverse reactions of white employees. On the other hand, he cites other situations, notably at Memphis, but also at Evansville, where strong union support by the union officials against wildcat stoppages designed to thwart the nondiscrimination policy bolstered the company's stand and assured Negroes of their rights.[46]

At the Louisville plant of Harvester, rival union situations complicated integration. The UAW won bargaining rights at the foundry, its later merged rival, the Farm Equipment Workers, won in the main assembly plant. In addition, a number of craft unions represented the skilled trades. In our discussion in subsequent chapters, we shall point out how such union fragmentation can inhibit intraplant movement and upgrading of minorities.

The other principal unions in the two industries are the United Steelworkers and the International Association of Machinists. The Steelworkers union has always admitted Negroes without discrimination, and afforded them their full contractual rights, but like most unions, it generally accepts management's employ-

---

[45] *Ibid.*, p. 191.

[46] Hope, *op. cit.*, pp. 100-115, 121-124.

ment policies. Thus there is no record of the Steelworkers challenging the concentration of Negroes in foundries and blast furnaces and their virtual exclusion from other departments such as rolling mills.[47] The IAM, originally a craft union, formerly excluded Negroes from membership by a provision in its ritual.[48] Now it admits Negroes without discrimination. We shall return for an up-to-date look at union racial policies after examining company policies in the 1960's.

---

[47] Northrup, *et al., op. cit.,* Part Four, pp. 330-334.

[48] Herbert R. Northrup, *Organized Labor and the Negro* (1944; reprint ed., New York: Kraus Reprint Co., 1971), pp. 3, 8. The exclusionist clause, which pledged members to recommend only "competent white mechanics" as members was deleted from the ritual in 1949.

CHAPTER IV

Negro Employment in the 1960's

The 1960's did not begin auspiciously for Negro employment in the farm equipment or construction machinery industries. In 1960, the former industry employed nearly 50,000 fewer employees than it had a decade earlier. Although Negroes did not lose jobs disproportionately, the census reported only 5,497 black farm equipment workers in 1960 as compared with 7,290 a decade earlier (see Table 8 above). The downturn in employment in 1960 was further reflected by the fact that the proportion of black workers employed by International Harvester stood at 9.3 percent, the lowest in nearly twenty years (see Table 9 above). Moreover, in one year, Harvester would close its pioneer Chicago McCormick works, which in 1959 had a work force that was 37.1 percent black.

Data for the construction machinery industry are not provided by the decennial census. The industry, however, was also hit by adverse business conditions in 1960. Undoubtedly, Negroes, being the most recently employed and concentrated in semiskilled and unskilled blue collar jobs, were seriously affected.

DEVELOPMENTS IN THE EARLY 1960's

In the early 1960's, the situation did not improve substantially despite energetic efforts on the part of the new Kennedy Administration and intensified pressure from Negro groups for more and better jobs. New jobs just did not materialize rapidly, and were lacking especially in areas in which Negroes were concentrated.

President Kennedy's Executive Order No. 10925 in 1961 differed most significantly from earlier presidential orders creating fair employment practice committees by requiring all companies contracting with the federal government to submit periodic compliance reports giving the number of Negroes employed and the types of jobs they filled, and by introducing the "affirmative action" concept. Most farm and construction machinery firms were covered by this order.

Also in 1961, the federal government asked the cooperation of business in Plans for Progress—a program for improving the employment situation of Negroes through voluntary company initiative. International Harvester, Caterpillar, Deere, Allis-Chalmers and others all signed soon after the program was announced, pledging to take "affirmative action" to ensure that their nondiscrimination policies were implemented. Some plans also included a promise to make a deliberate effort to recruit more Negro applicants. Although Plans for Progress has been subject to considerable criticism because performance has often not matched its public promise, the program is significant as a symbol of management consensus that equal employment opportunity should be publicly proclaimed as a company goal.

Company Response

For International Harvester, Plans for Progress was largely, as this writer has noted earlier, "a reaffirmation of the company policy of welcoming minority members to its employment and of providing them with equal opportunities for training and advancement in the Harvester plants." [49] Yet by 1964, Negro employment at Harvester stood at 9.4 percent of the total, almost identical to the 1960 proportion. The closing of the McCormick plant, with its high proportion of Negroes, apparently offset any gains made elsewhere in the company.

For other companies, more rigorous government support of equal employment and membership in Plans for Progress meant new and special affirmative action attempts to recruit Negroes and other minorities for the first time, and usually also, central staff monitoring to insure that the new policies were effectuated. Of course, some companies moved more quickly and energetically than others. For example, between February 1, 1963 and February 1, 1965, Caterpillar doubled its minority group work force from 637 to 1,278.[50] The bulk of the minority employees were Negroes, as shown in Table 12, but the total black proportion was only 2.5 percent despite the much higher proportion at the Joliet works.

Companies such as Deere, Allis-Chalmers, and Case all began affirmative action programs after signing Plans for Progress agreements, as did many smaller companies. For most com-

[49] Ozanne, *A Century of Labor-Management Relations, op. cit.*, p. 193.

[50] Habbe, *Company Experience with Negro Employment, op. cit.*, p. 104.

TABLE 12. *Caterpillar Tractor Company Minority Employment by Occupational Group February 1, 1965*

Occupational Group	Negro Employees	Other Nonwhite Employees
Officials and managers	1	5
Professionals	2	13
Technicians	4	5
Office and clerical	42	26
Total white collar	49	49
Craftsmen	38	45
Operatives	581	151
Laborers	285	26
Service workers	53	1
Total blue collar	957	223
Total	1,006	272

Reprinted from Stephen Habbe, *Company Experience with Negro Employment*, Personnel Policy Study No. 201 (New York: National Industrial Conference Board, 1966), Vol. I, p. 105, by permission of the publisher. © 1966 by the National Industrial Conference Board.

Note: Total employment: 39,511; percent Negro: 2.5; percent other nonwhite: 0.7.

panies, however, the proportion of Negroes remained low, particularly where the plants were located away from the centers of Negro population.

On the basis of data in our possession, we estimate that by 1964, Negro employment in the farm equipment industry had gained slightly over that in 1960, and was about 5.5 percent of total employment, as compared with 4.5 percent in 1960.[51]

For construction machinery we have data in 1964 for the first time. The sample, set forth in Tables 13 and 14, includes ap-

[51] We have 1964 data in our possession for 16 companies in the farm equipment industry employing 90,396 persons. Virtually all these employees, however, were found in two companies, International Harvester and Deere, with about 2,500 others in a tractor plant of Ford in Detroit that was 25 percent black. The unrepresentative character of the data showed a Negro employment proportion of 8.8 percent, obviously not consistent with the situation extant in 1964.

TABLE 13. *Construction and Related Machinery Industry Employment by Race, Sex, and Occupational Group 177 Establishments United States, 1964*

Occupational Group	All Employees			Male			Female		
	Total	Negro	Percent Negro	Total	Negro	Percent Negro	Total	Negro	Percent Negro
Officials and managers	6,036	5	0.1	5,977	5	0.1	59	—	—
Professionals	4,465	8	0.2	4,437	8	0.2	28	—	—
Technicians	4,748	16	0.3	4,567	16	0.4	181	—	—
Sales workers	2,602	1	*	2,499	1	*	103	—	—
Office and clerical	12,536	123	1.0	5,968	62	1.0	6,568	61	0.9
Total white collar	30,387	153	0.5	23,448	92	0.4	6,939	61	0.9
Craftsmen	24,453	582	2.4	24,388	582	2.4	65	—	—
Operatives	29,064	2,060	7.1	28,570	2,048	7.2	494	12	2.4
Laborers	6,074	913	15.0	5,930	911	15.4	144	2	1.4
Service workers	1,489	212	14.2	1,370	202	14.7	119	10	8.4
Total blue collar	61,080	3,767	6.2	60,258	3,743	6.2	822	24	2.9
Total	91,467	3,920	4.3	83,706	3,835	4.6	7,761	85	1.1

Source: Data in author's possession.

* Less than 0.05 percent.

TABLE 14. *Construction and Related Machinery Industry Percent Distribution of Employees by Race and Sex 177 Establishments United States, 1964*

Occupational Group	All Employees Total	All Employees Negro	Male Total	Male Negro	Female Total	Female Negro
Officials and managers	6.6	0.1	7.1	0.1	0.8	—
Professionals	4.9	0.2	5.3	0.2	0.4	—
Technicians	5.2	0.4	5.5	0.4	2.3	—
Sales workers	2.8	*	3.0	*	1.3	—
Office and clerical	13.7	3.2	7.1	1.7	84.6	71.8
Total white collar	33.2	3.9	28.0	2.4	89.4	71.8
Craftsmen	26.7	14.8	29.2	15.2	0.8	—
Operatives	31.8	52.6	34.1	53.4	6.4	14.1
Laborers	6.7	23.3	7.1	23.7	1.9	2.3
Service workers	1.6	5.4	1.6	5.3	1.5	11.8
Total blue collar	66.8	96.1	72.0	97.6	10.6	28.2
Total	100.0	100.0	100.0	100.0	100.0	100.0

Source: Computed from Table 13.

* Less than 0.05 percent.

proximately 39 percent of the total industry's employment in 1964 (see Table 3 above) and is based on a wide geographical and size distribution of companies. The proportion of Negroes—4.3 percent—is believed, based upon field investigations and discussions, to be a quite accurate representation of the situation one year prior to July 1965 when Title VII of the Civil Rights Act of 1964 became effective.

Table 14 shows that nearly all Negro employees in construction machinery in 1964—96.1 percent—were concentrated in blue collar jobs, whereas only two-thirds of all employees were hourly workers. The Negro white collar workers were almost entirely female office and clerical employees. This indicates that office and clerical employees involve one group for which affirmative action can be prompt. The turnover is usually sufficiently high to provide openings, and unlike the blue collar areas generally, there are no recall lists under seniority arrangements which have to be exhausted pursuant to union agreements before new personnel can be employed.[52] Within the blue collar group in 1964, Negroes in construction machinery were overconcentrated in the bottom three categories and underrepresented in the craftsmen category. Over 80 percent of all the blacks were found in the bottom three categories as compared with about 40 percent of all employees. In 1964, it was obvious that Negroes were underrepresented in this industry, both quantitatively and qualitatively.

For farm equipment, comparable data are not, as noted, available. Based on the figures supplied by a few major concerns, however, we estimate that the racial-occupational distribution was not substantially different from that of construction machinery, although overall black employment was higher, largely as a result of the efforts of International Harvester.

AFFIRMATIVE ACTION AND EMPLOYMENT EXPANSION, 1966-1969

Employment in the construction machinery industry rose steadily between 1961 and 1969, gaining some 94,000 persons in the process, an increase of almost 50 percent. Farm equipment employment rose less and declined earlier, reaching a peak of 142,200 in 1967 from a 1961 low of 108,800, and then declining to 135,400 by 1969 (Table 3). A labor shortage at the time of this expansion, combined with strong civil rights pressures, led to improvement in the black workers' position in both industries. For reasons that our discussion will make clear, however, the end of the decade saw both industries having a black worker complement decidedly less than such industries as automobiles and steel.

Large Company Affirmative Action, 1965-1970

Overall data for the two industries were not available prior to 1969, but trends are discernible by examining the figures for several large concerns combined. Table 15 shows the annual

[52] Because of the heavy layoffs in the late 1950's, recall lists in some plants and industries were not exhausted until 1963.

TABLE 15. *Farm Equipment and Construction Machinery Industries, Annual Changes in Employment by Race Four Companies, 1965-1970*

Year	Total Employment	Negro Employment	Percent Negro
1965	174,867	10,799	6.2
1966	180,200	11,157	6.2
1967	192,349	12,360	6.4
1968	181,512	12,062	6.6
1969	n.a.	n.a.	n.a.
1970	179,015	13,153	7.3

Source: Data in author's possession.

changes in total and black employment between 1965 and 1970 for four major concerns.

During this six-year period while total employment in these industries rose only 2.4 percent, Negro employment rose 21.8 percent. Yet this progress toward integration moved the Negro percentage of total employment from 6.2 percent in 1965 to only 7.3 percent in 1970. The low overall level of Negro employment makes even Herculean efforts at the employment office in a short period seem insignificant. For example, at one firm during a one-year period, 59 percent of the new employees hired were Negro. This increased its Negro employment 20 percent, yet moved its overall percentage of Negroes only from 3.2 to 3.8 percent. But to have a significant impact on the proportion of Negroes employed, this firm and others must keep up their present hiring policies for a ten-year period. That they well might be doing so is indicated by the fact that total employment declined by 13,334 between 1967 and 1970, but during this period Negro employment, after first declining, increased by 793.

The Process of Negro Advancement

While Table 15 shows the overall trend in Negro employment between 1965 and 1970 it does not show in what occupational groups the Negro gains were made. Tables 16 and 17 show Negro employment proportions and distributions by occupational groups for five large firms over a six-year period, 1964 to 1970.

TABLE 16. *Farm Equipment and Construction Machinery Industries Employment by Race, Occupational Group, and Percent Change Five Large Companies, 1964 and 1970*

Occupational Group	1964			1970			Percent Change 1964 to 1970	
	Total	Negro	Percent Negro	Total	Negro	Percent Negro	Total	Negro
Officials and managers	14,941	21	0.1	18,273	120	0.7	22.3	471.4
Professionals	9,275	20	0.2	10,980	84	0.8	18.4	320.0
Technicians	7,563	52	0.7	8,166	93	1.1	8.0	78.8
Sales workers	2,962	3	0.1	2,843	13	0.5	—4.0	333.3
Office and clerical	25,485	290	1.1	25,721	904	3.5	0.9	211.7
Total white collar	60,226	386	0.6	65,983	1,214	1.8	9.6	214.5
Craftsmen	30,215	1,015	3.4	31,733	1,575	5.0	5.0	55.2
Operatives	76,567	7,704	10.1	70,755	8,221	11.6	—7.6	6.7
Laborers	13,134	1,834	14.0	10,832	1,971	18.2	—17.5	7.5
Service workers	3,149	468	14.9	6,376	718	11.3	102.5	53.4
Total blue collar	123,065	11,021	9.0	119,696	12,485	10.4	—2.7	13.3
Total	183,291	11,407	6.2	185,679	13,699	7.4	1.3	20.1

Source: Data in author's possession.

TABLE 17. *Farm Equipment and Construction Machinery Industries, Percent Distribution of Employment by Race and Occupational Group Five Large Companies United States, 1964 and 1970*

Occupational Group	1964 Total	1964 Negro	1970 Total	1970 Negro
Officials and managers	8.2	0.2	9.8	0.9
Professionals	5.1	0.2	5.9	0.6
Technicians	4.1	0.5	4.4	0.7
Sales workers	1.6	*	1.5	0.1
Office and clerical	13.9	2.5	13.9	6.6
Total white collar	32.9	3.4	35.5	8.9
Craftsmen	16.5	8.9	17.1	11.5
Operatives	41.8	67.5	38.1	60.0
Laborers	7.1	16.1	5.9	14.4
Service workers	1.7	4.1	3.4	5.2
Total blue collar	67.1	96.6	64.5	91.1
Total	100.0	100.0	100.0	100.0

Source: Computed from Table 16.

* Less than 0.05 percent.

Table 16 also shows percentage change over the period. These five firms employ 58 percent of the workers in the two industries. None of the five have less than 1,000 employees.

In Table 16 we see that the total number of jobs in the segment of these two industries represented rose by only 2,388 or 1.3 percent from 1964 to 1970. Yet in the same six-year period white collar jobs increased by 5,757 or 9.6 percent. Of these new white collar jobs the whites picked up the overwhelming number, 4,929, and the Negroes gained 828.

In the blue collar jobs the Negroes did much better relative to whites. Here the total number of jobs declined by 3,369 or 2.7 percent, but Negroes gained 1,464 blue collar jobs. Thus on balance the whites "abandoned" 4,833 blue collar jobs while picking up 4,929 white collar jobs, a net job gain of 96 or 0.06 percent.

The Negroes gained in both categories, 828 white collar and 1,464 blue collar jobs, a net gain of 2,292 or 20.1 percent. In purely economic terms this was a good trade for both groups, assuming that many of the Negroes finding jobs in these high paying firms came from lower paying jobs, from high schools, and from the unemployed. This is a significant economic gain for Negroes no matter how regrettable it is that in absolute terms the gain was largely in the blue collar field.

An optimist might call attention to the high percentage gains of Negroes in white collar employment, 214.5 percent over the base year compared to a 13.3 percent gain over the base year in blue collar employment. However, it is the extremely low absolute levels of Negro employment in white collar jobs in the base period that permit small absolute gains to be translated into extremely high percentage gains. For example, while the number of Negroes in official and management positions increased from 21 to 120, a gain of 471.4 percent, Negroes still held only 120 out of 18,273 or 0.7 percent of the jobs in this category. They made little entry into the sales field. Their absolute white collar gains have been in the clerical division.

In the blue collar field Table 16 indicates absolute employment gains in every category. This is an encouraging sign since it indicates upgrading of Negroes and entry of Negroes into the higher paying jobs of craftsmen and operatives. Negro gains as craftsmen look impressive but are misleading, because, as mentioned earlier, the designation "craftsmen," covers such a broad category of jobs that it conceals the fact that the Negro penetration into truly skilled trades has been almost negligible.

Furthermore, the future for this category is not bright because of the almost universal failure to enroll a sufficient number of Negroes in the apprenticeships. Although not enough data were available on the race of apprentices to include this group in Table 16, some significant data are available. In 1964 several companies included in this study had no Negro apprentices. By 1968 several reported only one Negro apprentice each. The company with the best record in the industry had only six percent Negro apprentices, fifty percent below its Negro percentage for craftsmen. On the basis of incomplete data on Negro apprenticeship, the two-industrywide percentage of Negro apprentices is estimated at not over two percent of total apprenticeships.

THE OVERALL PICTURE, 1969

By 1969, the high point of the long prosperity of the 1960's was over. Employment in the construction machinery industry reached an all time peak in that year, and declined somewhat in 1970; but the farm equipment industry reached its peak employment in 1967 and declined thereafter (Table 3). Data for 1969, therefore, reflect a peak employment period for the one industry, but a period of some unemployment for the other.

Data for the two industries covering companies with 100 or more employees are set forth in Tables 18 and 19, as supplied by the Equal Employment Opportunity Commission. About 80 percent of the employment in the two industries is included in these data, with employment in all regions and major locations well represented.

Table 18 again shows evidence of continued expansion of black employment in the two industries. In farm equipment, the black employment ratio was the highest noted so far—6.8 percent—as compared with a census figure of 4.2 percent in 1960 and an estimated 5.5 percent in 1964. Construction machinery, an industry expanding faster and longer during the decade than farm equipment, increased its proportion of Negro employees to 6.9 percent in 1969 as compared with 4.3 percent in 1964. Thus a combination of employment expansion and civil rights pressures resulted in the construction machinery industry increasing its black employment at a substantially faster rate than did the farm equipment industry and, by 1969, having a slightly higher percentage as well.

In terms of occupational distribution, the industries differed little in 1969, both from each other and from data previously presented. Table 19 shows that for both industries, about 94 percent of the blacks were in blue collar positions as compared with about two-thirds of all employees. This, to be sure, represents an improvement over the situation noted for previous years, but clearly indicates the long road ahead before equality can be achieved. The following analysis by occupational categories will highlight the problems involved in attempting to integrate the work force in these industries.

Officials and Managers

The officials and managers category includes all management personnel from foremen to company president. In 1969, only

TABLE 18. *Farm Equipment and Construction Machinery Industries Employment by Race and Occupational Group 864 Establishments United States, 1969*

Occupational Group	Combined Industries			Farm Equipment			Construction Machinery		
	Total	Negro	Percent Negro	Total	Negro	Percent Negro	Total	Negro	Percent Negro
Officials and mangers	29,271	156	0.5	9,857	54	0.5	19,414	102	0.5
Professionals	15,981	85	0.5	4,604	22	0.5	11,377	63	0.6
Technicians	17,213	203	1.2	4,946	54	1.1	12,267	149	1.2
Sales workers	8,225	10	0.1	1,910	—	—	6,315	10	0.2
Office and clerical	41,099	916	2.2	12,198	281	2.3	28,901	635	2.2
Total white collar	111,789	1,370	1.2	33,515	411	1.2	78,274	959	1.2
Craftsmen	71,197	3,464	4.9	14,121	398	2.8	57,076	3,066	5.4
Operatives	119,320	12,640	10.6	41,001	3,719	9.1	78,319	8,921	11.4
Laborers	30,489	5,156	16.9	13,282	2,432	18.3	17,207	2,724	15.8
Service workers	9,216	837	9.1	4,788	314	6.6	4,428	523	11.8
Total blue collar	230,222	22,097	9.6	73,192	6,863	9.4	157,030	15,234	9.7
Total	342,011	23,467	6.9	106,707	7,274	6.8	235,304	16,193	6.9

Source: U.S. Equal Employment Opportunity Commission, 1969.

Note: Data represent 229 farm equipment and 635 construction machinery establishments.

TABLE 19. *Farm Equipment and Construction Machinery Industries*
Percent Distribution of Employment by Race
864 Establishments
United States, 1969

Occupational Group	Combined Industries Total	Combined Industries Negro	Farm Equipment Total	Farm Equipment Negro	Construction Machinery Total	Construction Machinery Negro
Officials and managers	8.6	0.7	9.3	0.7	8.3	0.6
Professionals	4.7	0.3	4.3	0.3	4.8	0.4
Technicians	5.0	0.9	4.6	0.7	5.2	0.9
Sales workers	2.4	*	1.8	—	2.7	0.1
Office and clerical	12.0	3.9	11.4	3.9	12.3	3.9
Total white collar	32.7	5.8	31.4	5.6	33.3	5.9
Craftsmen	20.8	14.7	13.2	5.5	24.2	19.0
Operatives	34.9	53.9	38.4	51.1	33.3	55.1
Laborers	8.9	22.0	12.5	33.4	7.3	16.8
Service workers	2.7	3.6	4.5	4.4	1.9	3.2
Total blue collar	67.3	94.2	68.6	94.4	66.7	94.1
Total	100.0	100.0	100.0	100.0	100.0	100.0

Source: Computed from Table 18.

* Less than 0.05 percent.

156 Negroes held such positions in 864 establishments reporting to the EEOC. In both farm equipment and construction machinery, they comprised 0.5 percent of the total, in farm equipment, Negro officials and managers comprised 0.7 percent of all black employees; in construction machinery, 0.6 percent. It is, moreover, very likely that most Negroes so classified are first line supervisors. In addition, most of the large companies have appointed equal employment opportunity coordinators who are usually black. This position is attached to the personnel department and has provided Negroes not only with opportunities for good jobs, but also jobs from which promotion to line and staff positions of general responsibility have occurred.

Traditionally, management positions in these industries, as in others, are won by those who work up the line through the marketing, production, or finance functions. As we shall note shortly, marketing has been almost closed to blacks. Production personnel usually have engineering training. We shall note also below that few Negroes have been, or are taking training in this profession. When they are, they seem to remain or be confined to research and design.[53] There is also a great shortage of black accounting and financial personnel.

Lower and middle management jobs are also filled by those who work up from the lower ranks. The low proportion of Negroes in the two industries reduced the potential for such upgrading until recently. Another management source has been the graduate business schools, which had few Negro students in the past. Now this is changing, but all industry is competing for this talent which is still relatively scarce.

Data in our possession for individual companies show substantial percentage increases between 1964 and 1969 in the number of black officials and managers; in terms of numbers, however, as Tables 18 and 19 show, the gains have not changed the overall minute representation of blacks in such jobs.

Professionals and Technicians

The farm equipment and construction machinery industries are not highly technical ones, if one judges that by the percentage of the work force employed as professionals or technicians. Table 19 shows that less than 10 percent of the work force was classified in these categories. For Negroes, there was great underrepresentation: they held only 0.5 percent of the professional jobs and 1.2 percent of those classified as technicians. Only 1.2 percent of all black employees were found in these categories taken together.

Most professionals in these industries are engineers, although there are the usual number of accountants, medical doctors, etc. Likewise, most technical positions are in work that is supportive to engineering. Very few Negroes were trained as engineers in the past, primarily because this was an area where few could obtain jobs. Now jobs are indeed available. Thus Dr. Robert Kiehl, who has been carefully analyzing the supply of and demand for black engineers, reported in October 1970 that "There seems to be

---

[53] Robert Kiehl, *Opportunities for Blacks in the Profession of Engineering* (Newark, N.J.: Foundation for the Advancement of Graduate Study in Engineering, 1970), p. 8.

no question but that there are widespread education and employment opportunities for blacks in engineering and in technicians' work." But he also found that only "About 2 percent of all engineering students in the U.S. are black. The percentage of blacks in Bachelor of Technology programs is about 5 percent as is also the approximate percentage of blacks in technicians programs." [54]

Dr. Kiehl found that about one-half of the Negro engineering students were in six predominantly black colleges, all but one located in the South, and the one in Washington, D.C. He concluded that "relative lack of information on engineering coupled with employment discrimination of the past seem to be the chief reasons for the apparent lack of interest of blacks in this profession today." [55]

In other professions, exclusive of the traditional ones of ministry, education, and law, Negroes are also in very short supply. For our two industries, however, the number and percentage of blacks in professional and technical jobs is low even granting that. Undoubtedly, the location factor discussed below is significant. A black professional in a town where few Negroes dwell and fewer still black professionals are found is in a lonely spot. Given the demand by companies in other areas, such personnel would likely be difficult to recruit. Still the numbers and percentages, as will be noted later in this chapter, do not compare favorably with other industries.

Sales Workers

If one were to designate the occupational group most inhospitable to black employment, one would unhesitatingly choose sales. In 1969, the construction machinery industry employed ten black sales workers—0.2 percent of its sales work force. This was ten more than were employed by the farm equipment industry.

There are the usual reasons. Selling is often a highly social activity, with family entertaining, access to private clubs and golf courses, etc., required. Companies fear the adverse reaction of potential customers. Given the few Negroes that dwell in agricultural areas, and the antipathy often found in the construction industry relating to Negroes in better jobs, company

---

[54] *Ibid.*, p. 14.

[55] *Ibid.*

misgivings are understandable (although not necessarily supportable). Finally companies point to the need for many salesmen to be technically trained and the shortage of blacks with such training.

Despite these problems, many companies have begun to utilize black sales personnel. If minorities are to be integrated fully into our society, they cannot be left out of key industrial functions. In some industries, the social and customer problems have been found to be less serious than feared.[56] As to the technical training, the first requirement is to employ more blacks so qualified. Dr. Kiehl's study found that no black engineers surveyed were in sales jobs.[57] Our two industries are in this respect following a general practice in industry.

Office and Clerical

Office and clerical work is an area where companies in these two industries, as in many others, have obtained relatively quick results in their attempts to expand black employment. Nevertheless, the results have not yielded an outstanding success—only 2.2 percent of the overall clerical labor force (2.3 percent for farm equipment) was black in 1969. In the five large firm sample (Table 16), office and clerical had reached 3.5 percent by 1970, perhaps indicating a better job-integration effort by these large firms. To be sure, the companies in these industries have a number of problems—location being a key one—which make recruitment of black clerical personnel difficult. Nevertheless, the numbers and percentages remain low even when these factors are considered.

Main Office Employment

An examination of data relating to main office white collar employment throws light on the problems and policies of the industries relating to white collar employment. Such data are in the author's possession for five companies for 1964 and 1968.

Company A initially had an extremely low percentage of Negro white collar employment, 0.6 percent. After four years and a

---

[56] See, for example, F. Marion Fletcher, *The Negro in the Drug Manufacturing Industry,* The Racial Policies of American Industry, Report No. 21 (Philadelphia: Industrial Research Unit, Wharton School of Finance and Commerce, University of Pennsylvania, 1970), pp. 51-58.

[57] Kiehl, *op. cit.,* p. 8.

35 percent increase in total office employment, Negro employment increased only to 0.8 percent. This cannot be said to represent an all-out effort at employment integration. In Company B with total office employment practically stable, the Negro employment ratio rose from 1.1 percent to 1.6 percent. As in Company A the greatest increase occurred in the "office and clerical" category. Ninety-three percent of the Negroes in the main office remained in this category during both periods. Although this is the greatest success achieved among the five companies, it still is minimal.

Company C with no Negroes in its office in either period apparently may not have taken employment integration in the office seriously by 1968. There is one mitigating factor, however, in the case of Company C. In the "office and clerical" category which is where gains were registered in Companies A and B, Company C had a 40 percent decline in total employment. In Company D, Negro employment in 1968 was limited to the clerical category, there being no other white collar integration. Company E, although having a very low overall Negro ratio of 1.1 percent in the office, had the highest percentage of its Negro employees in the professional category—0.6 percent in 1968.

In summary, although two firms increased the proportion of Negroes in the office from the 1964-65 to the 1967-68 period, the initial levels were so low that the increases are symbolic only. When one pinpoints these tiny increases they are almost entirely in "office and clerical," the lowest white collar category. There does not seem to be much relationship between a firm's equal opportunity efforts in the plant and in the office. In the plants, as we shall see, real progress has been made. In the offices Negro employment is generally only tokenism. Firms in communities of high Negro population have failed to do much better in the proportion of Negroes hired for their offices than firms in communities of low Negro population. This probably indicates that until recently the main offices actually have been neglected in the firms' equal opportunity policy.

Even where company reports reveal a fair number of Negroes in the white collar category, the actual jobs of Negroes listed as "white collar" by companies may not be in the offices, but in the plants as inspectors and minor supervisors, semi-blue collar jobs. For example, in 1968 one firm reported 83 Negro white collar employees in the entire company. Yet only 12 of these 83 were found in the main office, 7 of them in "clerical." Another firm reported 38 Negro "officials and managers" but only 2 of them are in the main office, the balance are supervisors.

Although this author believes that the slim progress revealed by data on main office integration warrants the criticism levelled above, such criticisms must be modified by a realistic appraisal of the limited supply of professionally, technically, and clerically trained Negroes. As a result of past discrimination, societal as well as company, there is an appalling shortage of Negroes trained for white collar occupations. With Negro white collar employment in the main offices at such low levels as late as 1964, it is clear that the effort by all firms to raise it suddenly to an appropriate level in accordance with government directives would be seriously hindered by supply shortages. We have already commented on this fact. Even though in many cases industry has lowered its employment standards and greatly increased its in-service training programs, the number of trained Negroes needed to meet the increased demand is nowhere in sight.

Craftsmen

Table 18 shows that 5.4 percent of the construction machinery craftsmen are black, almost double the ratio in farm equipment. Table 19 appears to accentuate the better showing of the former industry. Thus, 19 percent of all black construction machinery employees were craftsmen in 1969 as compared with 24.2 percent of all employees; by contrast, the percentages for farm equipment were 5.5 percent for black employees and 13.2 percent for all employees. These industrywide data of Tables 19 and 20 imply that there are proportionately more craftsmen in construction machinery than in farm equipment and that the construction machinery industry is advancing Negroes into craftsmen's jobs more rapidly. However, the much higher percentage of craftsmen, both all employees and Negro, reported for the construction machinery industry as compared with farm equipment may involve possible classification problems as mentioned earlier.

Furthermore, despite the overall higher proportion of craftsmen and of blacks in the craftsmen category in the construction machinery industry, a comparison of the big mass production firms in the two industries does not reveal any such differences. The higher proportion of craftsmen in construction machinery is in part caused by the larger proportion of moderate sized companies and the "job shop" nature of the specialty work done by them. The number of items in a product line is often relatively small, and each may have special-order aspects. Mechanics are likely to assemble such vehicles or equipment, using helpers who work up to craftsman status. Negroes employed as helpers work up to higher status in this manner.

In plants visited by the author or by other Industrial Research Unit personnel (and these were larger companies), it was found generally that Negro craftsmen tended to be concentrated in the foundries in such crafts as molders and coremakers, rather than being broadly distributed throughout the plant. Those craftsmen outside of the foundries were generally in such trades as millwright, painter, or carpenter but were considerably underrepresented in the electrical, plumbing, and mechanical trades. This is typical of American industry generally. Unfortunately, these are the trades commanding the highest wage rates which are primarily in demand in metal working industries such as the two studied here.

As in similar industries, the farm equipment and construction machinery industries draw their craftsmen from three sources: employment of those who learned the skills elsewhere, including immigrants; company apprentice programs; and upgraders or learners trained on the job after production experience. As Dr. Northrup has pointed out in regard to the similar situation in the automobile industry:

> The apprentice programs have always attracted a high type person—the high school graduate with good mechanical and mathematical aptitude who did not make it to college because of family attitudes or finances. They have traditionally been the "middle class" of automobile workers—a white elite who did not associate with the production men. Like building tradesmen, the skilled trades group have been antagonistic toward accepting Negroes into their group. Moreover, supervisory personnel in the skilled trades area have traditionally come up through the ranks, and undoubtedly share the general attitudes attributed to the rank and file.[58]

Most employees who work as craftsmen have not served apprenticeships, although the proportion with apprentice training is higher in the electrical and mechanical trades than in those such as carpentry, painting, or the trowel trades (bricklaying, plastering, cement finishing, etc.). Some "pick up" the trades by working with relatives or friends; others are white immigrants

---

[58] Northrup *et al.*, *Negro Employment in Basic Industry*, Part Two, p. 81.

who learned their trades abroad. Since the shortage of Negro craftsmen is universal, attracting such employees does not increase the proportion of Negroes in these industries' craft categories.

There are formidable obstacles in gaining Negro craftsmen through company apprentice programs. Historically, with good reason, Negroes have regarded both the training and the jobs in the skilled trades, particularly the electrical, plumbing and mechanical ones, as closed to them. Even when apprentice training becomes available, many Negroes are deterred by the fear of entering a probably still hostile white world.

Poor schooling in southern segregated and northern inner city schools has resulted in a substantially higher percentage of apprentice examination failures among Negro applicants than among whites. Probably, Negroes who can master the reading comprehension and mathematics examinations to qualify for apprenticeship programs prefer to go to college if it is possible. In 1969 International Harvester at Memphis initiated a big recruitment drive to get young Negroes to enter skilled trades apprenticeships. Broad community-wide efforts were made to get applicants to write the examinations. The effort was a disappointment. Only a handful of Negroes wrote the examinations. None of them passed.

In addition, the wages of the apprentice can be less than those of an unskilled laborer in the foundry or assembly line. Many Negroes cannot afford to accept present lower wages for the sake of future financial gain.

Beginning in 1967, the Ford Motor Company has sought ways to break down the barriers impeding blacks from becoming apprentices. The pioneer in opening apprenticeships to Negroes, Ford always has had a relatively high proportion of black applicants but they too have had difficulty overcoming educational disabilities. As part of this program, journeymen's jobs were studied to determine the relationship of training and related education to actual work needs and those who failed pre-apprentice examinations were given special tutoring and retested. The results have broadened the potential source of supply of apprentices and expanded the Negro share of such jobs.[59]

International Harvester, Caterpillar and three firms in other industries are each beginning similar programs which, like

---

[59] From interviews conducted by Industrial Research Unit staff.

Ford's, are especially designed to open up apprentice training to minority employees by improving their capacity to handle the related education requirements of mathematics and reading comprehension, and also by making the job instruction fit carefully the future work required of journeymen. Such a flexible approach could well result in a greater share of black craftsmen in trades where they have heretofore been largely unrepresented.

Probably, the most likely means of increasing the proportion of Negro craftsmen is through the upgrading of Negroes already employed as operatives in the plants. Many companies have special training programs for this, but training in itself is not enough. To be in a craft department a worker must meet certain minimum performance standards. Regardless of his training, he can do this only with the cooperation of his fellow workers and foremen in the department. Thus the black entering a craft department is at the mercy of the workers and foremen who may all be white and who may simply resist integration. Another obstacle to the promotion of a Negro operative to a craft department is departmental seniority which will be discussed in the next chapter.

Operatives

Operatives include entry jobs and also those which require varied training from very little to considerable before proficiency is achieved. In this classification, the two industries have done their best integration jobs. Given the labor force and population mix where the plants are concentrated, Negroes have a reasonably fair share of these jobs—10.6 percent overall in 1969 (Table 18).

Actually, of course, Negroes are overconcentrated in operatives jobs. Table 19 shows that 53.9 percent of all blacks are operatives, but only 34.9 percent of all employees are so classified. Moreover, within the operatives classification, we observed a definite tendency for blacks to be overconcentrated at the lower ranges.

Nevertheless, these are entry jobs and as such, are the means by which newcomers work into industry. They pay well and provide some opportunities for upgrading. Integrating this occupational group is the first step, and this the industries herein involved have seemingly accomplished.

Laborers and Service Workers

Only 11.6 percent of all employees in the two industries are classified in the two lowest blue collar categories but 25.6 percent of the Negro employees are found there. These are often deadend jobs although upgrading from them does and has occurred. They include janitors, yardmen, porters, cafeteria workers, and in some plants, material movers. Since unionization in the 1930's, the term "laborer" has been applied to a smaller segment of industry. This is partly because the term is considered unpalatable to the individual, partly because mechanization and automation have altered the character of the work. It is typical of industry, of course, for Negroes to be overconcentrated in such jobs.

Female Employment

Tables 20 and 21 show female employment for the two industries by occupational group. As we noted in Table 5, only 9.6 percent of the two industries' workers in 1970 were female. The great bulk of these are office and clerical employees, some 70.8 percent of all women being in this occupational group. Most of the rest are in operative and laborer jobs. The industries do not have large numbers of light assembly work for which females are employed in many industries.

For Negro women, the occupational distribution is significantly different. Only 24.2 percent of the black females are in office and clerical jobs, but 24.3 percent work as operatives and 45.4 percent as laborers. We must conclude that despite the improvement noted in the employment of Negro women in office and clerical work since 1960, the industry remains inhospitable to using women in better jobs, and beyond that, black women are in a markedly inferior position to white women in terms of jobs held.

LOCATION FACTORS

The concentration of the farm equipment and construction machinery industries in the Midwest has already been noted. In most other studies in the series, regional racial employment comparisons have been made, but for these two industries they would have very limited usefulness. The three largest producers have plants in the West and there are a few in the South. The overwhelming concentration in the Midwest means that any re-

TABLE 20. *Farm Equipment and Construction Machinery Industries*
Female Employment by Race and Occupational Group
864 Establishments
United States, 1969

Occupational Group	Combined Industries			Farm Equipment			Construction Machinery		
	Total	Negro	Percent Negro	Total	Negro	Percent Negro	Total	Negro	Percent Negro
Officials and managers	299	2	0.7	98	—	—	201	2	1.0
Professionals	437	5	1.1	105	—	—	332	5	1.5
Technicians	712	23	3.2	285	11	3.9	427	12	2.8
Sales workers	199	1	0.5	42	—	—	157	1	0.6
Office and clerical	24,641	465	1.9	7,499	160	2.1	17,142	305	1.8
Total white collar	26,288	496	1.9	8,029	171	2.1	18,259	325	1.8
Craftsmen	525	26	5.0	50	1	2.0	475	25	5.3
Operatives	4,475	467	10.4	1,742	204	11.7	2,733	263	9.6
Laborers	2,910	872	30.0	2,064	793	38.4	846	79	9.3
Service workers	604	61	10.1	195	10	5.1	409	51	12.5
Total blue collar	8,514	1,426	16.7	4,051	1,008	24.9	4,463	418	9.4
Total	34,802	1,922	5.5	12,080	1,179	9.8	22,722	743	3.3

Source: U.S. Equal Employment Opportunity Commission, 1969.

Note: Data represent 229 farm equipment, 635 construction machinery establishments.

TABLE 21. *Farm Equipment and Construction Machinery Industries*
Percent Distribution of Female Employment by Race
864 Establishments
United States, 1969

	Combined Industries		Farm Equipment		Construction Machinery	
Occupational Group	Total	Negro	Total	Negro	Total	Negro
Officials and managers	0.9	0.1	0.8	—	0.9	0.3
Professionals	1.2	0.3	0.9	—	1.5	0.7
Technicians	2.0	1.2	2.4	0.9	1.9	1.6
Sales workers	0.6	*	0.3	—	0.7	0.1
Office and clerical	70.8	24.2	62.1	13.6	75.4	41.0
Total white collar	75.5	25.8	66.5	14.5	80.4	43.7
Craftsmen	1.5	1.3	0.4	0.1	2.1	3.4
Operatives	12.9	24.3	14.4	17.3	12.0	35.4
Laborers	8.4	45.4	17.1	67.3	3.7	10.6
Service workers	1.7	3.2	1.6	0.8	1.8	6.9
Total blue collar	24.5	74.2	33.5	85.5	19.6	56.3
Total	100.0	100.0	100.0	100.0	100.0	100.0

Source: Computed from Table 20.

* Less than 0.05 percent.

gional analysis would be comparing the great bulk of the industry with a few scattered plants. We shall, however, as we have in earlier sectors, refer to individual plants located outside, as well as within, the Midwest in discussing various factors affecting black employment.

The impact of location on Negro employment is not only seen in the rural Midwest location of these two industries, but it can be observed also within a city merely by measuring the closeness of a plant to Negro population concentrations. This is demonstrated by Figure 3 in which seven plants of seven different companies located in Milwaukee, Wisconsin are charted in relation to their percentage of black employment and the plant's distance from the Negro housing district.

FIGURE 3. *Farm Equipment and Construction Machinery Industries, Percent Negro Employment by Distance from Negro District Seven Establishments, Milwaukee, 1968*

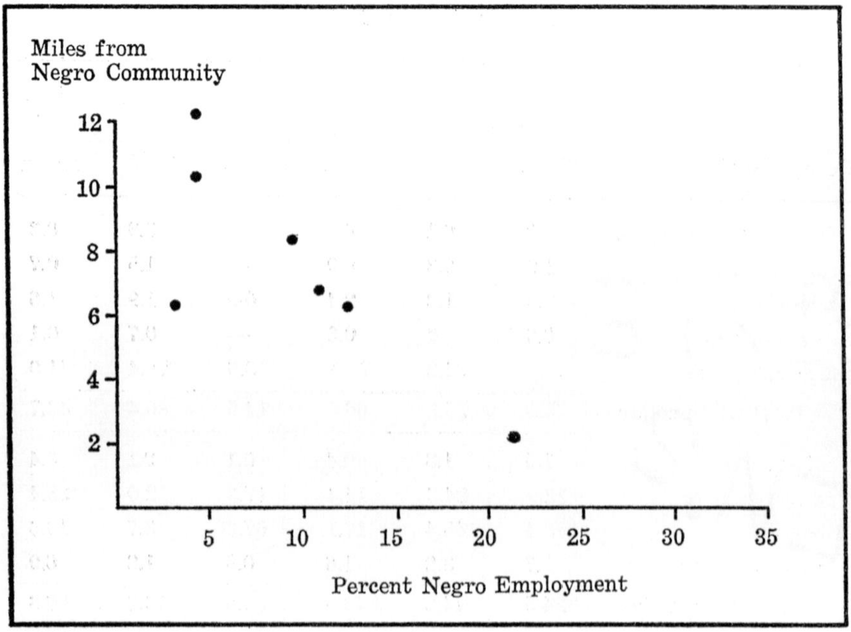

Source: Data in the author's possession.

In this city the highest percentage of Negro employment occurred in the plant closest to the Negro district. The plants with the lowest percentage of Negroes had two geographical similarities. They were quite distant from the Negro residential district and were located in ethnic neighborhoods. Clearly, in these seven plants the nearness of a plant to the Negro community would appear to be the most important factor in Negro employment. Conversely, there might be the additional factor of the location of other plants in neighborhoods which the Negroes viewed as hostile to them.

The closing of the older International Harvester plants in Chicago has had a decidedly adverse impact on Negro employment. The area around the older plants, as in many cities, had become heavily black populated. This was particularly true in regard to the plants closed most recently—McCormick in 1961, 37 percent black, and Tractor in 1971, 35 percent black. These plants have not been replaced. Their production has been dispersed to newer plants in the suburbs or in different parts of the country. In either case, they are away from concentrations of Negro population.

In these plant closings company policy and understandings with the UAW give workers in closed plants a job preference over prospective new hires at other Harvester plants. A transferred worker can carry with him his seniority back to 1961. But an exception was apparently made recently when, at the request of the United Automobile Workers, International Harvester exempted the Hough plant from the obligation to accept the 1961 seniority of workers hired from a closed plant.

Hough is a material-moving vehicle plant located in a heavily white westside suburb of Chicago. A young plant, it had been purchased by Harvester in the mid-1960's. It was nonunion at the time of the Harvester acquisition. Its workers averaged far fewer years of service than those at Tractor already tagged for closing. Vainly the UAW sought to win representation rights at Hough through NLRB elections. Hough workers feared that had they voted UAW they might have had to accept Tractor transferrees who had more seniority than they did, even though Tractor workers could carry with them only those years from 1961 onward. The fact that many Tractor workers were black allegedly heightened the reluctance of Hough employees to vote UAW and thus become subject to the union-management agreement. The "black" issue was, of course, never given official cognizance by the UAW nor by International Harvester.

After the loss of two NLRB elections the UAW requested that Harvester suspend the policy permitting employees to carry seniority if offered a job at Hough. Harvester consented. The UAW won the next election. Tractor workers could carry with them up to 9 years of seniority if accepted for a job at Chicago plants such as Melrose Park or West Pullman but only one year of seniority at Hough.

Thus two conflicting pressures have affected black employment at Harvester. On the one hand, the company has been striving to increase the proportion of Negroes in its plants and offices while, on the other hand, it has been moving an increasing proportion of its production to the suburbs and to other outlying locations which causes a decrease in the proportion of Negro employees.

Caterpillar's plants also strongly reflect locational patterns. Its highest black percentages are in the Joliet and Oakland, California plants, both of which have sizable black populations nearby. In the Peoria area, the central office is close to the small Negro population center; the East Peoria plant is reasonably nearby, but across the Illinois River. Newer plants, however, are located in outlying towns and areas which have very few Negroes. These plants reflect the character of the labor force.

Much the same can be said for all other plants in the industry—those of Deere, Case, and many others. One can almost look at their location and give the overall black complement. There is, of course, variation from plant to plant and company to company, but the location factor is strong throughout.

COMPARISON WITH OTHER INDUSTRIES

An examination of how the farm equipment and construction machinery industries compare to related industries casts additional light on their performance and problems in the area of Negro employment. Both industries are part of the broad two digit classification, non-electrical machinery (SIC 35). Table 22 compares the overall non-electrical machinery data with those in the two industries studied here. Table 23 compares the two industries with five other major industries. Table 24 compares the two industries with the chemical and electrical manufacturing industries.

Looking first at Table 22, we find that both the industries studied here have a higher overall black percentage than did the larger group. For farm equipment, this higher percentage is attributable to its greater concentration of Negro operatives and laborers; in the other categories, including each white collar occupational group except one and that of craftsmen, percentage of Negroes in farm equipment is below that of the large industry group of which it is a segment. The only exception is officials and managers in which both farm equipment and non-electrical machinery have 0.5 percent black representation. Obviously, the 1969 data show the farm equipment industry to have fallen short in the upgrading of blacks and their utilization of them in better jobs as compared with similar industries, but to have done better in overall employment.

The construction machinery industry data reveal an identical pattern, with one significant exception. The overall representation of Negroes is superior to the non-electrical machinery, but the white collar representation is inferior. The craftsmen's ratio

TABLE 22. *Non-Electrical Machinery, Farm Equipment and Construction Machinery Industries Employment by Race and Occupational Group United States, 1969*

Occupational Group	Non-Electrical Machinery (SIC 35)			Farm Equipment (SIC 352)			Construction Machinery (SIC 353)		
	Total	Negro	Percent Negro	Total	Negro	Percent Negro	Total	Negro	Percent Negro
Officials and managers	117,512	580	0.5	9,857	54	0.5	19,414	102	0.5
Professionals	96,159	1,043	1.1	4,604	22	0.5	11,377	63	0.6
Technicians	87,588	1,413	1.6	4,946	54	1.1	12,267	149	1.2
Sales workers	34,691	98	0.3	1,910	—	—	6,315	10	0.2
Office and clerical	175,388	4,516	2.6	12,198	281	2.3	28,901	635	2.2
Total white collar	511,338	7,650	1.5	33,515	411	1.2	78,274	959	1.2
Craftsmen	297,859	10,296	3.5	14,121	398	2.8	57,076	3,066	5.4
Operatives	448,951	40,315	9.0	41,001	3,719	9.1	78,319	8,921	11.4
Laborers	110,503	16,462	14.9	13,282	2,432	18.3	17,207	2,724	15.8
Service workers	25,533	3,458	13.5	4,788	314	6.6	4,428	523	11.8
Total blue collar	882,846	70,531	8.0	73,192	6,863	9.4	157,030	15,234	9.7
Total	1,394,184	78,181	5.6	106,707	7,274	6.8	235,304	16,193	6.9

Source: U.S. Equal Employment Opportunity Commission, 1969.

for construction machinery is, however, superior. Assuming the validity of these data, this represents considerable progress. The probable reasons for the higher percentage have been discussed earlier. The percentage of black craftsmen in construction machinery is one of the highest in major industries, as the comparison in Table 23 will show.

Table 23 compares the percent black employment in five major industries with that in farm equipment and construction machinery. The data for the five industries are for 1968; those for farm equipment and construction machinery for 1969. Since for the industries generally, 1969 data were likely to show a higher proportion of black employment than for previous years because of continued affirmative action, the data in Table 23 should be biased in favor of the two industries studied here.

Looking at the data in Table 23, we find that the overall proportion of Negroes in farm equipment and construction machinery is considerably less than that in automobile, steel, and rubber tires, but exceeds that in aerospace and petroleum. The three industries which have the highest black ratio are, interestingly enough, very similar in occupational content to each other and to the two studied here—that is, two-thirds to three-fourths of all employees are blue collar and the largest occupational group is the semiskilled operative classification. Thus the farm equipment and construction machinery industries have a smaller percentage of blacks than do three similarly structured industries. We believe that location is the principal reason for this.

The fact that the proportion of Negroes in aerospace and petroleum is less than that of farm equipment and construction machinery is largely a matter of occupational structure. The dominant labor force requirements for the former two industries, and especially for aerospace, are for professional, technical, and skilled personnel. The shortage of such personnel among black people greatly reduces the proportion of Negroes in the labor force.[60] In effect, occupational and skill requirements offset locational advantages, other things being equal.

In terms of the proportion of salaried and craftsmen jobs that Negroes hold, the record of the two industries studied here is again poor, with the overall white collar percentage less than any

[60] For a discussion of the aerospace and petroleum industry racial policies, see Northrup *et al., op. cit.,* Parts Three and Six.

TABLE 23. *Seven Manufacturing Industries*
Percent Negro Employment by Occupational Group
United States, 1968

Occupational Group	Automobiles	Aerospace	Steel	Rubber Tires	Petroleum	Farm Equipment[a]	Construction Machinery[a]
Officials and managers	1.4	0.6	1.3	0.9	0.3	0.5	0.5
Professionals	0.7	0.9	0.5	0.8	0.7	0.5	0.6
Technicians	1.5	2.1	2.2	2.1	1.7	1.1	1.2
Sales workers	b	0.6	0.3	5.0[c]	2.7	—	0.2
Office and clerical	4.4	3.4	2.9	1.9	4.8	2.3	2.2
Total white collar	2.4	1.7	1.9	1.6	2.4	1.2	1.2
Craftsmen	3.3	5.4	6.3	2.8	2.2	2.8	5.4
Operatives	21.3	13.5	18.7	12.8	7.3	9.1	11.4
Laborers	29.1	20.7	28.7	22.8	18.0	18.3	15.8
Service workers	27.3	23.2	20.0	29.2	21.5	6.6	11.8
Total blue collar	18.6	10.0	16.7	11.7	6.2	9.4	9.7
Total	14.5	5.4	13.3	8.3	3.9	6.8	6.9

Source: Herbert R. Northrup, Richard L. Rowan, *et al.*, *Negro Employment in Basic Industry*, Studies of Negro Employment, Vol. I (Philadelphia: Industrial Research Unit, Wharton School of Finance and Commerce, University of Pennsylvania, 1970), Part Eight, p. 727; and U.S. Equal Employment Opportunity Commission, 1969.

[a] 1969 data.

[b] Few sales workers combined with office and clerical.

[c] Including many non-sales personnel; actually very few Negroes in this classification.

of the other five industries. Again the one bright spot is the high percentage of black craftsmen in the construction industry which exceeds that of all industries except steel.[61]

Our third and final interindustry comparison is set forth in Table 24 in which data for the combined farm equipment and construction machinery industries are included with those of the chemical and electrical manufacturing industries for 1969. These are both industries which require a higher percentage of salaried employment than do farm equipment or construction machinery. Both also, however, use a large proportion of semiskilled workers and craftsmen.

Table 24 shows again that farm equipment and construction machinery have a smaller black ratio, and in addition, are employing a smaller proportion of Negroes in all white collar classifications than the other two industries. Again the bright spot is craftsmen, where the construction machinery data put our two industries slightly ahead.

Industry Comparisons—Final Comment

These industry comparisons must be evaluated with care. In previous studies in this series, we have pointed out the variety of factors which affect an industry's racial policies, and many of these are not within a company's direct control.[62] Nevertheless, the consistent pattern revealed here does emphasize two factors: the significance of location which contributes substantially to the low black utilization in the two industries; and the slowness of the industry to employ Negroes in salaried positions. The former, to be sure, affects the latter, as we have noted. Nevertheless, white collar utilization is so consistently below patterns in other industries that we can reconfirm our earlier statement to the effect that affirmative action, at least until quite recently, has been concentrated in the plants.

---

[61] For an analysis of the reasons for the high percentage of black craftsmen in steel, see *ibid.*, Part Eight, pp. 728-730.

[62] See *ibid.*, Part Eight for a discussion of these factors.

TABLE 24. *Three Industry Comparison Employment by Race and Occupational Group United States, 1969*

Occupational Group	Electrical Manufacturing (SIC 36)			Chemical and Allied Products (SIC 28)			Farm Equipment and Construction Machinery (SIC 352 and SIC 353)		
	Total	Negro	Percent Negro	Total	Negro	Percent Negro	Total	Negro	Percent Negro
Officials and managers	137,731	1,265	0.9	110,177	934	0.8	29,271	156	0.5
Professionals	176,851	1,844	1.0	91,802	1,233	1.3	15,981	85	0.5
Technicians	124,664	3,921	3.1	58,319	2,395	4.1	17,213	203	1.2
Sales workers	24,438	158	0.6	49,142	489	1.0	8,225	10	0.1
Office and clerical	211,348	8,243	3.9	123,395	4,928	4.0	41,099	916	2.2
Total white collar	675,032	15,431	2.3	432,835	9,979	2.3	111,789	1,370	1.2
Craftsmen	210,937	8,861	4.2	141,657	6,614	4.7	71,197	3,464	4.9
Operatives	664,749	76,593	11.5	252,102	32,770	13.0	119,320	12,640	10.6
Laborers	169,545	21,616	12.7	80,490	18,283	22.7	30,489	5,156	16.9
Service workers	27,411	5,253	19.2	22,539	5,294	23.5	9,216	837	9.1
Total blue collar	1,072,642	112,323	10.5	496,788	62,961	12.7	230,222	22,097	9.6
Total	1,747,674	127,754	7.3	929,623	72,940	7.8	342,011	23,467	6.9

Source: U.S. Equal Employment Opportunity Commission, 1969.

CHAPTER V

Impact of Company, Union, and Government Policies

In some other studies in this series, we have compared the overall data for the industry with those of the major companies. The overriding importance of the locational factors in racial employment matters in the farm equipment and construction machinery industries result in data that are not very meaningful if the largest company figures are combined for such purposes. We shall, therefore, discuss here those efforts by the larger companies which have attempted to increase their minority employment within the broad framework of factors that have already been found to affect the extent and nature of such employment.

In turn, these efforts have been affected by union and government policies and by the changing mix of population, housing, labor market, and other factors discussed in the previous chapter. This chapter discusses individual company policies, and union and government impacts thereon.

THE 1967-1968 STUDY OF ALLIS-CHALMERS

In 1967-1968, Allis-Chalmers provided historical and current data which gave greater insight both into the progress which is being made toward equal employment opportunity and the practical difficulties which are encountered in such a course. This company has long been a community leader in employment integration. Among Milwaukee Negroes it has a favorable image as a place to work, a very important asset in attracting Negro applicants. The company has pioneered in employing high school counselors during the summer to acquaint them with job opportunities. It has set up "hard core" employment programs and summer jobs for Negro youth.

In the 27-year period, 1941-1968, the percentage of Negro employment at the Allis-Chalmers West Allis Works rose from 0.7 percent to nearly 8 percent while the percentage of Negroes in the metropolitan area of Milwaukee grew from 2 percent to 14.1 percent. The occupational and departmental location of Negroes at the West Allis Works, as of February 1967, can be summarized as follows: almost one-half the Negroes were employed in the foundry; the rest were concentrated in the semiskilled operative jobs on the various assembly lines. Of a total of 886 Negroes in the combined plant and main office, only 23 were in the main office which had some 5,600 employees. Despite all the years of effort to upgrade Negroes, this total of 23 Negro office employees was but five more than the number of Orientals who comprised less than 1/10 of one percent of the population in the Milwaukee area. Of these 23 Negro office employees, 14 were clerks, 3 professionals, and 6 technicians. Looked at in another way, of all the salaried employees in both plant and main office, roughly 12,000 employees, only 40 were Negroes. There were 2 (or 3) foremen, 25 clerks and key punch operators, 4 technicians, 4 draftsmen, 1 interviewer, 2 engineers, 1 chemist, and 1 physician.[63]

This low white collar complement was the situation despite strenuous efforts on the part of the company to increase both its white and blue collar black representation. Table 25 demonstrates this policy. It shows that in the salaried field, 6 percent of the

TABLE 25. *Allis-Chalmers Manufacturing Company Percent Employees, Applicants, and Hirees by Race West Allis Works, 1967*

	Salaried			Blue Collar		
	Employees	Applicants	Hirees	Employees	Applicants	Hirees
White	98.7%	92.0%	93.0%	84%	70%	74%
Negro	0.7	7.5	6.0	14	30	24

Source: Allis-Chalmers Manufacturing Company.

Note: Balance to 100 percent other races or unknown.

[63] These data were provided by Allis-Chalmers Manufacturing Company, as were others relating to the company in this chapter.

hires were Negro compared with a present Negro salaried employment ratio of only 0.7 percent. In the blue collar field, 24 percent of the hires were Negroes compared with an employment ratio of only 14 percent. (Later figures for the first 6 months of 1968 provided by the company raise the Negro blue collar hires from 24 percent to 34 percent.)

At first glance this looks encouraging and would seem to portend a rapidly rising proportion of Negro employment at Allis-Chalmers. However, the proportion of Negro employees in the total work force is determined not only by the ratio of hires, but also by the termination rates.

In early 1968 the hourly Negro termination rate was 8 percent compared to 3 percent for whites. The overwhelming number of terminations was voluntary. For salaried terminations the rate was 25 percent for Negroes and 5 percent for whites. Generally in industry terminations are inversely related to length of service. Since the Negro has shorter service his quit rate is higher than the white, but not necessarily higher if equated for length of service. The high voluntary quit rate among salaried Negroes may look discouraging to Allis-Chalmers, but it may also reflect the very tight job market for trained Negroes. If we assume that the salaried quits are the result of better job offers elsewhere, a high quit rate is an encouraging sign in the total problem. The West Allis facility is a great distance from the Negro district so that in a tight job market Negroes may leave Allis-Chalmers as jobs open up closer to their residences.

There are little data (none of it recent) with which to analyze blue collar quits. In 1948 Allis-Chalmers compiled for one month a comparison of white and Negro quits in the foundry by occupation. In the skilled jobs of "coremaker" and "molder" in which the overwhelming majority of the workers were white, the white quit rate exceeded the Negro. But in the less skilled "chipper and grinder," and "helper" classifications, the Negro quit rate exceeded the white rate. In the unskilled "laborer" classification, where three-fourths of the employees were Negro, white quit rates exceeded black. There is thus some slight evidence here that for manual occupations when Negroes achieve seniority and skilled jobs their job stability may be higher than that of whites. The higher overall quit rate of Negroes does mean that the hiring rate for Negroes must have been even higher in 1948 than today in order to have significantly raised the Negro employment ratio.

An index of the lower seniority held by Negroes is revealed by data on active and inactive employees classified by race. A September 1968 breakdown of employees at Allis-Chalmers on the active and inactive (or basically layoff) rolls shows the following: Negroes constituted 13.9 percent of the employed hourly workers, but 19.9 percent of the hourly employees on layoff. Negroes constituted 1.9 percent of the salaried employees, but 2.9 percent of the salaried employees on layoff. Plantwide, Negroes constituted 7.6 percent of the total actively employed workers, but 13.9 percent of those on layoff.

One of the somber, long-run hinderances to employment integration is the lack of trained, skilled Negroes. The largest and growing number of job opportunities at Allis-Chalmers, as in the country generally, is in the white collar field. Fifty-four percent of Allis-Chalmers' hiring in the first six months of 1968 was of salaried employees. It is here that Negroes held only 0.7 percent of the jobs and had a quit rate of 25 percent. It is apparent that improvement in employment integration at Allis-Chalmers must be achieved to a great extent in the white collar area. Low Negro employment representation in the white collar category is the result not of neglect, but of some years of deliberate effort to increase Negro employment in this area. Table 25 points out the crux of the problem. Only 7.5 percent of the applicants for salaried positions were Negro.

The data on Negro employment, application, hires, and quits at Allis-Chalmers gives a view of the problems which the comany faces in trying to develop a more integrated work force. Many of these problems cannot be solved by the company alone. Much of the future upgrading of Negroes must come from community efforts in improving education so that more Negroes can qualify for the white collar jobs, professional, managerial, and clerical and for the apprenticable craftsmen occupations such as machinist, patternmaker, and electrician.

DEERE, CASE, AND CATERPILLAR

The plants of Deere, Case, and Caterpillar all are located in areas in which relatively few Negroes dwell. Nevertheless, all three have developed programs designed to increase their percentages of black employees and to utilize such employees in other than traditional jobs. These programs are described briefly below.

Deere and Company

Deere, the leading producer of farm equipment, is headquartered in Moline, Illinois. Except for a Los Angeles facility, all domestic Deere plants are located in the Moline area, or in nearby midwest cities, principally Des Moines, Waterloo, and Ottumwa. The black population of these cities varies somewhat, but is in all cases close to or under 8 percent.[64]

We noted earlier that Deere's first black employees entered the company during World War II, and worked in the foundries. In World War II, Negro employment expanded somewhat, but was both small overall and still concentrated in foundries, unskilled and semiskilled jobs. In 1957, Deere appointed the first Negro supervisor in one of its quad-city factories, and in 1959, employed its first Negro woman salaried worker for its general office. In 1963, the company began recruiting at black colleges.[65]

Deere joined Plans for Progress in 1964, and thereafter initiated a number of affirmative action programs designed to improve its minority employment representation. In 1964, 3.4 percent of its employment was black.

As part of its program, Deere has taken the lead in organizing the Quad-City Merit Employment Council which encourages minorities to complete their schooling, gives them guidance and provides them with job opportunities in local plants. An equal opportunity brochure, describing the kinds of jobs blacks hold at Deere, was developed and widely distributed, Additionally, plant remedial education and upgrading programs have been made available, and hardcore and National Alliance of Businessmen's JOBS programs are included in the company's efforts. In terms of overall employment, results have been slow, with black employment still below 4 percent in 1970, and a total of 84 Negroes in salaried positions throughout the company in 1969.

J. I. Case

Case, now a subsidiary of Tenneco, is headquartered in Racine, Wisconsin, and has major facilities there, in Rockford and Rock Island, Illinois, Burlington and Bettendorf, Iowa, and Terre Haute, Indiana. Again, these are not cities with substantial black populations.

[64] U.S. Census of Population, 1970 advance reports.

[65] Based on information supplied by the company.

Starting much later than Allis-Chalmers or International Harvester, the Case Company has engaged in a broad variety of community and plant programs to end discrimination in both the community and on the job. In 1963 under its then President, Merrill D. Hill, Case initiated a School-Work Experience Program in Racine aimed at potential school drop-outs. Other leading employers have joined the program which offers one-half day of paid work experience and one-half day of schooling. The employer group assures each graduate a job at the end of the program.

Case admits that prior to 1963, its choice of white or Negro employees was dictated by the labor market conditions. Not until World War II were Negroes admitted even to foundry employment. At this time and during the Korean War they were heavily employed in the foundry.

From 1958 through 1961 employment demands at Case were modest. Moreover, in 1958 Massey-Harris shut down its Racine agricultural implement plant releasing onto the Racine market hundreds of experienced, skilled whites. Case hired many of them. A 1962 sharp boom in Case orders resulted in rapid expansion of employment. It was at this time that Negroes broke their restriction to the foundry and entered broadly into assembly departments. Since then the company has been working as a matter of conscious policy to achieve equal employment opportunity through hiring, training, and upgrading.

At Case progress in employment integration has been substantial since 1962. There are now several Negro foremen. The company has granted leaves of absence to several Negro employees to take executive positions in community agencies.

Negro employees recognize that there has been improvement at J. I. Case, but, of course, tell of much more to be done. One Negro described the changing situation in Racine and at Case as follows (his comments are paraphrased):

> When I got out of high school in 1956 I was offered only a custodial job by a leading Racine employer. I refused and later got into the foundry at Case where every black went. Our first "break" came in 1960 when callbacks after the strike were on the basis of plant-wide seniority. Some of us found that outside the foundry there were good paying jobs that were clean and you didn't have to break your back. Pretty soon they made me foreman, but that's where it ends. I left when I couldn't go any higher. Case holds you down to foreman.[66]

[66] Interview, 1968. To what extent this attitude is typical is, of course, unknown.

In fact, Case does have Negroes in salaried positions higher than supervisor. In 1970, Case had an overall black complement in excess of 7 percent, and at the Racine plant, better than 10 percent. Given the low area black population, this must be considered a better than average result for its employment program in terms of numbers. Moreover, it slightly exceeds the industry average.

The union at the Case Racine plant, UAW Local 180, selected its first Negro executive board member in 1950. Since then Negroes have been active in the leadership, reaching up through the vice-presidency of the local in 1968.

Caterpillar

Caterpillar, the largest of the construction machinery concerns, has engaged in a variety of affirmative action programs since the early 1960's. We have already noted its involvement in a program designed to expand the number of black craftsmen. It has also engaged in a variety of programs for recruiting blacks, employing and training hardcore unemployed, and upgrading already employed disadvantaged.

In Peoria, where Caterpillar is the major employer, its officers and managers have played leading roles awakening its citizens to areas of racial discrimination in the community. Caterpillar's public affairs manager, as chairman of the Mayor's Commission on Human Relations, arranged numerous bus tours for Peoria civic leaders and citizens to the blighted areas of the city. The tours highlighted all the tragic conditions accompanying acute racial discrimination in housing. Some of the results have included: changed policies by some of the real estate companies, a city council request to the State of Illinois for a state equal housing law, a local ordinance on equal housing, movement of at least a few Negroes out of the ghetto area and into white neighborhoods, stricter enforcement of the housing codes, and adoption of an equal housing opportunity policy by Caterpillar for the considerable number of homes it markets as a result of facilitating employee transfers.[67]

We noted in Chapter IV that Caterpillar began an expanded program of seeking black employees after joining Plans for Prog-

[67] Byron DeHaan, "Enlisting Community Support for Civil Rights," speech before the National Industrial Conference Board, January 20, 1966 (mimeograph).

ress. This program began from a very low base of black employment, 1.5 percent of total domestic employment. By mid-1970, the percentage of Negroes on Caterpillar's payrolls was 5.5 percent, more than double the 1964 ratio, but still below the industry average.

Such comparisons raise the question of what should be considered a good performance under given circumstances. We have usually compared Negro employment with Negro representation in the population in this and related studies. Two factors, however, limit the validity of such comparisons. The first is that the black population is, outside of the South, generally concentrated in a particular segment of a city, while a large company such as Caterpillar draws its labor force from a wide area, especially when it has a number of plants which are located not only in the immediate urban area, but also in several outlying communities in which few Negroes dwell. Hence the potential labor force of the plants is based on a population that is much more heavily white than is that of Peoria alone.

But population and labor force racial proportions are not identical because of a smaller black labor force participation rate. The following interesting information was assembled by Caterpillar's personnel manager.

> The Peoria SMSA consists of Peoria, Tazewell, and Woodford Counties. According to Illinois State Employment Service reports, the population of these three counties is 352,300—of which 13,387, or 3.8%, are Negro. The three-county work force is 153,750—of which 5,350, or 3.5%, are Negro. It follows then that 5,350 of the 13,387 Negro population, or *40.0%*, are classed as being in the work force; 153,750 of the 352,300 total population, or 43.6%, are classed as work force; leaving 148,400 of the 338,913 white population, or 43.8%, as the white work force.
>
> We then applied the work force concept to the Caterpillar employment. We employ a total of 25,526 people from these three counties in our plants located in these same counties. Of these, 1,147 or 4.5% are Negro compared with a population percent of *3.8* and a work force percent of 3.5. A final computation shows that we employ 16.6% of total three-county work force, 16.4% of the white work force, and 21.4% of the Negro work force.[68]

Thus Negro employment at Caterpillar compares favorably with Negro work force representation and with Negro proportions in the larger area which Caterpillar's workers actually represent. Such comparisons for other companies might yield similar results.

---

[68] Letter dated April 22, 1970.

HARVESTER—OVERALL AND IN THE SOUTH

By special programs, such as hardcore employment, International Harvester has attempted to continue its leadership in equal employment. The problems have been severe. As noted, employment in the industry has declined, forcing cutbacks in newly hired employees and thus offsetting some special training and employment programs. In addition, the closing of older plants in the Chicago area has offset gains of Negroes elsewhere. By 1970, Harvester's black employment stood at approximately 11 percent of the company's total, down from the all-time high of 14.5 percent in 1955, but up from the 1960 figure of 9.3 percent.

The Memphis Plant Revisited—1969

A re-examination of the degree of employment integration in 1969 affords a significant test of what a committed management and a committed international union can and cannot accomplish in a southern city in the space of 20 years' time. Of course, during the last 10 years external national forces have aided the company and the union. The Kennedy and Johnson Executive Orders required substantial affirmative action toward integration for eligibility for defense contracts. The Civil Rights Act of 1964 set up the EEOC which provided an opportunity completely outside labor and management channels for minority redress of discriminatory practices. And, during this decade, the Negroes waged a civil rights revolution.

In spite of the work toward equal job opportunity by the companies, the union, Negroes, and the government, there is still a long way to go. As mentioned earlier, the Negro employment percentage was only 23 percent at the end of 1968, which is less than in 1953 and far short of 38.9 percent which is the percentage of Negro population in the city of Memphis. Despite John Hope's data on the advancement of Negro workers written in 1953, the plant is still characterized by lily-white and overwhelmingly black departments. Of the roughly 300 men in the truly skilled trades, there was but one Negro, an electrician who was on layoff in January 1969 because he had only 50 days' seniority in the electrical department. In welding, in 1969, there were roughly 3 Negroes out of 100, none of them with substantial seniority. In the machine department there were only 6 Negro machine operators out of 75 operators and inspectors, the most senior Negro having only six months' seniority. Among 279 foremen, 3 were Negro, 2 of them appointed in 1967, and the first in 1965. There were no Negro apprentices. Of 450 workers in the foundry, approximately 325 were Negroes, concentrated as usual in the hottest places, the forge shop and pouring the molten metal. Although the UAW Local 988 went through its intraunion battle to desegregate washrooms in the new union hall from 1958 to 1960, the plant washrooms and cafeteria remained segregated for several years thereafter. De facto segregation of washrooms adjacent to predominantly white departments continued until very recently.

The above situation in Memphis may be what International Harvester had in mind when its employee relations manager wrote in 1965:

> Although International Harvester was unquestionably among the early leaders in this area, its activities during the early years did not eliminate or, in some areas, even significantly reduce discriminatory employment practices. For example, it was not until a few years ago that the last vestiges of segregated facilities were removed from the last of several Southern facilities. Also there are a few operations where, while Negroes are employed in substantial numbers, yet, somewhat inexplicably, they are not employed in certain work areas or departments. More importantly, their absence cannot be easily explained away solely by something such as the obvious lack of qualified candidates.[69]

However, on the positive side, according to union officials there has been slow but steady progress since early 1960 and in the last three years the rate of progress has been stepped up. Union officials attribute the early achievement of some 21 to 26 percent of Negro employees in the late 1940's and early 1950's to the fact that in this early period the foundry was the largest and most important unit in the plant. During the 1950's, as the plant expanded, it was the assembly operations which grew. For this work the company hired more whites than Negroes. Since the early 1960's the company has been hiring Negroes and whites on about a 50-50 ratio. In 1969 there were Negro clericals in the office, very unusual for the South. Whites as well as Negroes are now hired at entry level jobs in the foundry. Total employment, however, has not been increasing much and therefore the opportunities for Negroes have not been as great as the com-

---

[69] Harry C. Baker, "A Voluntary Approach to Equal Opportunity," in Herbert R. Northrup and Richard L. Rowan (editors), *The Negro and Employment Opportunity* (Ann Arbor: Bureau of Industrial Relations, University of Michigan, 1965), p. 114.

pany's hiring ratio might imply. The actual increase in the percentage of Negroes employed is made possible by a higher turnover rate for whites.[70]

Improvement in the local attitudes and actions has been noticeable during this period. We have seen that prior to 1958, in spite of the insistence and direct action for integration by the UAW national leaders, the local union leadership, dominated by segregationists, colluded with foremen and other local company officials to slow down the Negro integration. Since the end of the Detroit-imposed administratorship of the local during 1959 and 1960, there has been a touch and go political battle within Local 988 between those forces which want to create a truly integrated local and those which at least wish to go slow. The latter includes members who believe in segregation. Of the 1969 union executive board of 16 members 10 were elected on a plant-wide basis and six were elected by zone. All elected to the plant-wide posts were white. Five of the six representing zones were white. The one Negro executive board member represented the foundry and the forge shop. The president and the chairman of the bargaining committee were white. Although the elected officials were, with one exception, white, a coalition of Negroes and whites favoring strict contract enforcement and equal treatment of whites and Negroes has held a slight but growing majority since 1961.

We thus, not surprisingly, have a picture of continued progress, but one that still is far from equality. Perhaps the Harvester personnel official who wrote the following in 1965 would still find it apt:

> Why do some of these practices continue to exist? For one thing, the company's policies are administered by people whose attitudes and emotions are pre-conditioned by environment, cultural background, and education. Also, the implementation of company policy in any area, certainly this one, is especially difficult where the communication of top policy must be disseminated to thousands of management people who represent us at the employee and community level. Has there been some discrimination at International Harvester under some of these policies? I am sure there has been—some due to pressures and influences of internal matters, some because of human failings, and others by external forces beyond our control. While we know that the overwhelming majority of our employees agree with the approach to this problem,

[70] The employment information quoted here was provided by union officers. They are estimates based on their intimate knowledge of the work force—not data based on payroll records.

there are always small pockets of resistence in any company concerning any program—particularly any change from the "status quo." Nevertheless, in the final analysis, the employer must provide the momentum for the elimination of discriminatory practices. It is a responsibility which cannot be delegated. Neither can it be conditioned upon its popular appeal.[71]

Despite its shortcomings, Harvester remains a leader in Memphis, in the South, and in its industry. A black employee interviewed by the author in Memphis in 1969 noted that the company had received a plaque for its leadership in equal employment in 1965. He first said, after complaining of alleged inequities: "I wanted to spit on that plaque." Then he added, "If I have to work here in the South in Memphis, let me work at International Harvester." [72]

A SOUTHERN COMPANY

The Memphis plant of International Harvester represents the racial employment policies of a progressive northern company applied to a southern plant unionized by an organization sympathetic to these policies. One way to note its accomplishments is to compare it with another deep South plant. One such plant, manufacturing large construction equipment, was visited. Table 26 shows the percentage of black employment, 1964-1968, the changes in total employment for the period, and the white and black employment occupational distribution. What is apparent is clearly racial exclusion from all white collar departments and low black utilization.

This company was awarded a defense contract which caused the great increase in total employment and in Negro employment. This increase took place only in the defense work of the company. In the traditional peacetime departments, the proportion of Negroes employed remained very low. There were no Negro foremen in this southern plant. In 1968 there were no Negro white collar workers. Subsequently, the firm hired its first Negro white collar worker, a secretary, and put her in the most visible place, the personnel department.

[71] *Ibid.*, p. 115.

[72] Interview, 1969.

TABLE 26. *Farm Equipment and Construction Machinery Industries*
Percent Negro Employment and Occupational Distribution by Race
One Southern Company, 1964 and 1968

Occupational Group	Percent Negro Employment		Percent Change in Total Employment	Occupational Distribution			
				White		Negro	
	1964	1968	1964-1968	1964	1968	1964	1968
Officials and managers	—	—	69.6	5.6	6.4	—	—
Professionals	—	—	63.8	4.1	4.6	—	—
Technicians	—	—	43.1	4.6	4.3	—	—
Sales workers	—	—	7.4	3.9	2.8	—	—
Office and clerical	—	—	43.8	6.2	6.1	—	—
Total white collar	—	—	47.2	24.4	24.2	—	—
Craftsmen	2.0	5.8	30.1	53.1	45.0	34.1	14.9
Operatives	7.1	26.8	145.6	16.5	21.7	40.9	42.8
Laborers	9.6	47.6	208.5	6.0	7.2	20.5	35.6
Service workers	100.0	40.0	3,150.0	—	1.9	4.5	6.7
Total blue collar	4.0	19.7	77.0	75.6	75.8	100.0	100.0
Total	3.0	15.7	69.9	100.0	100.0	100.0	100.0

Source: Data in author's possession.

In the absence of a union, seniority did not control job transfers or promotion. There was no posting of job vacancies, which were thus known only to management personnel and spread to nonsupervisory personnel by the "grapevine." With an all-white management it appeared that Negroes were not tuned in on "the grapevine." There was no job evaluation system and no systematic training of employees for advancement. Thus in the absence of an affirmative company policy or union established seniority, the Negro's opportunities for promotion were considerably less than in a situation like Harvester's Memphis plant.

HARDCORE EMPLOYMENT

Because of its significance as the largest concern in the two industries and the pioneer employer of Negroes, this study has dealt extensively with the policies of International Harvester. We have also noted that Harvester is involved in a program designed to increase the number of black craftsmen. Although there is much to be done to increase Negro mobility within the plant, the primary need in integrating employment, of course, is to hire many more Negroes. Several companies involved in this study have made a concerted effort to do this by establishing a special training program for the "hardcore" unemployed. Harvester is one of them and another prefers to remain anonymous.[73]

In April 1968, International Harvester began such a program at five plants in the Chicago area and at five plants in other cities. The original goal of 260 trainees was later expanded. Financed in part by a grant from the U.S. Department of Labor, the training was divided into two phases: four weeks of pre-job training at a Harvester training center outside any plant, followed by eight weeks of on-the-job training in the various plants.

The first problem in such a program was selection. International Harvester dropped all mechanical aptitude and mental tests. Only a medical examination was required. Applicants were not rejected for police records, nor for poor work records. The Woodlawn Community Organization and the Urban League assisted in locating candidates. The United Auto Workers Union, which has its own pre-employment training program, submitted

[73] Both programs discussed here are based on information supplied by the companies involved.

candidates and cooperated in the entire program. The Illinois State Employment Service certified only that the applicants were "hard core."

It would appear that the four-week training center is an important ingredient of the excellent retention rate which followed. Of the 145 men hired as of August 1968, only 17 dropped out within one year. None of these 145 men in the program had high school diplomas; many of them were members of street gangs. The cooperating organizations picked up the candidates in a company station wagon and took them to the orientation center or training center. Frequently friends were recruited in order to reinforce one another in the new experience. The company served breakfast during the first week. Non-job orientation included personal hygiene, buying on credit, banking, proper diet, use of the time clock, and union-management relations. Since no one was rejected for lack of reading or arithmetic skills, remedial courses in these subjects were instituted during the training program. Mechanical training began with simple sub-assemblies, pallet building, use of hand tools and materials, handling equipment, and packaging and shipping. Hardcore recruits earned $2.80 per hour during the orientation program and most started at $3.06 an hour when they moved into the 8-week training program at the plant.

The two biggest hazards and causes of loss of workers were police trouble and layoffs due to unemployment. In December 1968, of 200 hardcore recruits who had completed the four-week program and started on the regular plant work program, 26 were at that time on layoff. Despite the elaborate and apparently successful training program, despite the federal grant and the extensive cooperation by Negro organizations and the United Auto Workers, nothing had been done to mitigate the traditional plant layoff according to seniority: last in, first out. Further layoffs and the closing of the tractor plant greatly reduced the number still on Harvester's payroll. Nevertheless, the program is continuing and being expanded to other Harvester plants.

A second company, which asks to remain anonymous, instigated a hardcore training program and established a control group of normal hirees by which to measure the program's effectiveness. The hardcore group was selected in cooperation with minority organizations from persons certified as meeting "hard core" unemployment standards by the state employment agency. The control group was a control group only in that it was hired normally at the same time as the hardcore group. The groups were not matched as to education, age, nor minority status. Both groups began with the same entry level jobs and the same standard rate of pay (well over $3.00 per hour).

For the first four weeks, four hours a day were spent in the plant and the other four hours in a classroom. Reading, writing, and shop fundamentals constituted the classroom study with workers divided into two classes, those with less than four years of schooling and those with more than four years. For the 20 subsequent weeks the hardcore group spent two hours twice a week in the classroom. For their hours on the job, the trainees were widely scattered throughout the plant and not easily distinguishable from other employees, except for their absences to attend class or the presence from time to time of an instructor. Meanwhile, the control group went through only the normal fifteen hour factory training program plus the four to six weeks of on-the-job instruction for those who were promoted to machine jobs.

At the end of one year, the carefully kept records of both groups were compared. Retention rates were, of course, one of the interesting outcomes. One year after the hardcore group had begun training, 74 percent of the original 50 were still employed. Only 60 percent of the non-hardcore group were still employed. If the five non-hardcore employees on military leave were counted as actively employed, however, the retention rate for that group would be 70 percent. The difference between 70 percent and 74 percent is not statistically significant. The reasons for the loss or turnover were, however, different between the two groups. Thirteen, or 26 percent of the hardcore group had been discharged compared to only 7, or 14 percent of the control group. The primary cause for discharge among the hardcore group was rules violation (five), and poor attendance (five). Rules violations (four), were the main reason for discharge of the control group.

The other major reason for turnover of the control group was the voluntary quit. Eight of the control group quit. There were no voluntary quits among the hardcore employees. This lower ratio of quits among the latter was, of course, expected since their job marketability was much less. What is surprising, however, is that their retention rate was so high.

There were other important differences between the two groups. Thirty-two percent of the hardcore group received one or more warning notices against only 10 percent of the control group. The hardcore group lost a total of 418 days on medical leave compared to only 241 days lost for the control group. The control group had slightly more promotions and fewer demotions. Several months after the end of the training the absentee rate of the hardcore group was 9 percent compared with 6.2 percent for the control group.

Foremen were asked to rate employees regarding their capacity for promotion. On this rating, 77 percent of the control group employees were rated as promotable compared with 50 percent of the hardcore group. Foremen also rated employees as to their impression of the employee's ambition. On this rating 73 percent of the hardcore employees and 70 percent of the control employees were rated as ambitious. Although this may be a somewhat doubtful test, it at least indicates that so far as foremen are concerned, both had equivalently meritorious performance.

UNEMPLOYMENT AND AFFIRMATIVE ACTION

No matter how sincerely and effectively a company works to increase its black employment, the jobs must be available if its efforts are going to succeed. This is well indicated by the impact of the layoffs which occurred in the farm equipment and construction machinery industries beginning in 1968. For farm equipment, layoffs began in that year; for construction machinery, employment rose until 1969, then declined slightly.

Table 27 shows that in three firms Negro employment declined while total employment declined more slowly or actually rose slightly. This, of course, means that Negroes suffered substantially more job losses than whites. In these three firms, A, B, and C, among the Negroes it was the Negro blue collar worker who bore the brunt of unemployment, while the white collar Negro was favored over blue collar Negroes and indeed over white collar whites.

The cause of the blue collar Negroes' greater job loss is the operation of seniority under the union contract. Although seniority is color blind, it does require the layoff of the most recently hired. Since during the employment upswing of the sixties, firms A, B, and C had raised their proportion of black

TABLE 27. *Farm Equipment and Construction Machinery Industries, Percent Changes in Employment by Race and Job Class, Five Large Firms, United States, 1968-1970*

	Percent Change, 1968-1970				
Firm	Total Employment	Negro Employment	Negro Blue Collar Employment	Negro White Collar Employment	Total White Collar Employment
A	—17.2	—20.5	—23.5	+ 48.3	—16.0
B	—12.6	—15.0	—15.8	— 2.2	—12.1
C	+ 1.9	— 1.7	— 6.4	+ 60.2	+ 1.9
D	— 6.6	+ 3.2	+ 1.2	+ 35.7	— 6.0
E	+11.9	+62.2	+54.7	+150.0	+13.4

Source: Data in the author's possession.

employees over previous years, during the ensuing employment decline these same firms were required to lay off Negroes disproportionately.

Why did not the same thing occur among salaried employees? For one thing, they are not unionized and rigid seniority systems do not apply. In addition, there is a high attrition among female office workers because of marriage, pregnancy, husband moves, etc. Hence, these firms were able to continue to favor Negroes in hiring and increase their proportion of black white collar workers. The data indicate the possibility of whites being laid off and blacks hired, but this is an unrealistic assumption.

Firm D had a unique experience. It managed to increase absolutely its Negro employment, both blue collar (1.2 percent) and white collar (35.7 percent), while reducing total employment (6.6 percent). This means that so far as blue collar workers were concerned, either Negroes had slightly greater seniority in those plants where the job reduction occurred, or reductions occurred heavily in plants which were predominantly white. So far as D's white collar employees were concerned, it continued its affirmative action program by hiring proportionately more Negroes than whites and replacing white quits with Negroes.

In summary, it is clear that in three of the five firms listed in Table 27, the rate of forward progress toward equal employment opportunity of previous years for blue collar workers received a severe setback as a result of the employment decline of 1969-1970. Despite the general economic decline, four of the five firms continued to increase their number of Negro white collar workers. The fifth firm, B, although cutting its Negro white collar work force by 2.2 percent, actually increased its proportion of Negroes in these occupations since its total white collar force was cut by 12.1 percent.

Firm E increased its total employment and raised both Negro blue collar and Negro white collar employment at greater rates. All firms, those with total employment losses as well as gains, were able to make proportionately greater gains for Negro white collar workers. Although the small number and proportion of Negroes in white collar employment exaggerates the importance of these increases, it does indicate that affirmative action for salaried employees at least among the larger firms in this industry had begun to show some results by late 1970.

UNIONS, SENIORITY, AND INTRAPLANT MOBILITY

We have noted the role of unions and seniority a number of times in this and in previous chapters. Given the fact that the two industries are heavily unionized, and that seniority generally governs layoffs and largely also promotions within the bargaining unit, the impact of union policies, and particularly seniority, on the upward movement and job security of minorities is significant and worthy of detailed discussion.

Nondiscrimination Clauses in Union-Management Contracts

In the agricultural and construction machinery industries, the labor-management agreements contain no overt discrimination against minorities. On the contrary, practically every agreement includes a specific clause which prohibits both the union and the company from discriminating on the basis of race. The United Auto Workers took the lead in instituting these clauses long before the current civil rights upsurge. In the 1960's in the skilled trades, government, through the Department of Labor's Office of Federal Contract Compliance, put pressure on management and particularly on labor unions to add such clauses. For example, the last union at Allis-Chalmers to add such a clause was the union of maintenance electricians—the International Brotherhood of Electrical Workers—in 1968. In this case the company made the inclusion of such a clause a bargaining demand. This was rejected by the union. Government and company pressure finally brought union acceptance. An example of a strong nondiscrimination clause is found in the agreement between Deere and Company and the United Auto Workers. It reads as follows:

> There shall be no discrimination against any employee in the bargaining unit by either party under this agreement because of race, color, creed, religious belief, nationality, or sex. The company may take disciplinary action against any individual or groups of individuals violating the provision of the above sections.[74]

Unfortunately, the presence or absence of such clauses is largely a statement of policy which has litle actual impact on actual discrimination in the plant since lax enforcement and discriminatory hiring policies can negate the intent of the clauses.

Nature of Seniority

The purpose of seniority provisions is to insure that favoritism, or similar subjective factors, will not deprive persons of opportunities to retain their jobs or be promoted. Often seniority provisions are codifications or modifications of natural inplant job progressions. One moves from a less skilled to a more skilled job, using the former expenience to learn the latter. Seniority progressions can be by occupation, by department, by plant, or by some combination thereof. For Negroes, who may be concentrated in particular jobs or departments, or who may be recently employed, the widest application of seniority is beneficial if upgrading is the objective; if on the other hand, protection against layoff is the objective, the reverse is true. Wide seniority opens up the best jobs for all qualified workers in the plant; narrow seniority limits promotion to such jobs to a select group. But wide seniority permits workers throughout the plant to displace anyone with less seniority in the plant when layoffs occur.

Management usually prefers the narrower departmental or occupational seniority because it is less costly and simpler to operate. Narrow seniority can be carried out with fewer and simpler job transfers, and filling jobs from within the depart-

---

[74] All contract clauses are taken from labor-management agreements between the parties noted.

ment requires less expenditure on testing and training and less loss of production.

The skilled worker also is likely to prefer narrow seniority. He is inclined to want to have his own union or at least to keep the control over his job and wages out of the hands of the mass of less skilled workers in the production unit. In other words, he wants to establish a "property" right in his job and he does this by means of seniority. It would appear to be a reasonable quid pro quo that a worker who labors many years on a routine job as a helper have the first right to succeed to the next most skilled job above him. As these rights have been written into labor-management agreements, they do favor the workers who are geographically (in terms of the plant) and occupationally contiguous. A worker may feel outraged at having the prize job in his department go to a worker even in the same union but from a remote section of the plant. Yet such a system does tend to institutionalize the status quo, and therefore inhibit Negro advancement, as the following examples demonstrate.

Examples of Seniority Clauses

The impact of seniority may be better understood by examining examples of the several types of contract clauses. Departmental seniority for promotion is the most common in this industry. In the West Allis plant of Allis-Chalmers, the agreement between UAW Local 248 and the company describes departmental seniority for promotional transfers as follows:

> Preference in making a promotional transfer shall be given to the employees in the seniority department where the opening occurs. . . .

For transfers, a hierarchy is established:

> The applications for transfer which have been on file for more than 20 days will be considered in the following order of preference:
>
> First: Applicants from the seniority group in which the need has arisen; and
>
> Second: Applicants from the seniority division in which the need has arisen; and
>
> Third: Applicants from another seniority division.

Thus a person wishing to transfer from the foundry division to a different division and seniority group, such as the machine shop, would have little chance of accomplishing the transfer without accepting the bottom job in terms of pay in the seniority division to which he wished to transfer.

The UAW contract with Deere and Company covering a broad group of locals in nine plants likewise utilizes departmental seniority.

> Any employee who is working in the seniority classification and in the department in which the vacancy exists may request from the foreman of the department assignment to the vacancy. If more than one such employee requests assignment to the vacancy, the foreman will transfer the employee with the greatest seniority to the vacancy, provided he is qualified to perform the work. . . . In such event, the vacancy created by the transfer will be the vacancy to be filled by job bidding.

In the Bucyrus-Erie-United Steelworkers Agreement, occupational seniority, which is only slightly less restrictive than departmental seniority, is very simply stated:

> Senior qualified employees in the occupation in which the vacancy exists shall be assigned, upon request, to the vacant job without posting.

In contrast, plantwide seniority permits workers from any section of the plant to bid for any job in any other section of the plant, limited, of course, by one's ability to perform the job requested. The clause on "Upgrading and Transfer Procedures" in the contract between International Harvester's Farmall plant, Rock Island, Illinois, and the UAW provides an example:

> Where an employee is qualified to perform the job, seniority shall govern all transfers within the bargaining unit. Seniority and qualification shall govern all upgrading within the bargaining unit.

Clearly, this plantwide type of contract permits much more freedom for workers to transfer from one area to the other, from one type of job to the other.

It is, of course, possible to have modified plant and departmental seniority. Such an example is the agreement between J. I. Case Company of Racine, Wisconsin, and UAW Local 180. In the contract signed January 1, 1967, the union asked to change the departmental seniority to a modified departmental seniority. Formerly all vacancies were filled by giving preference to persons in the department. Under the 1967 agreement, only the first job posted and the second job subsequently made vacant if a department member fills the first are offered with preference to workers within the department. Thereafter, the future vacancies within the department are made available on the basis of plantwide seniority. This agreement reads as follows:

> Concurrent with the posting of the job, the company will poll the employees within the classification currently working within the department and select the employee in accordance with the criteria . . . below.
>
> The second vacancy created within the classification in the department as a result of (1) above, will be filled in the same manner and on the same basis as set forth above.
>
> Any other vacancy beyond the second must be filled by job posting and considering any employee in the plant. This clause is provided so that on any particular vacancy the maximum number of moves will not exceed 3.

Under this revision, should a welding job become vacant, another welder in the same department would have the first call on the job. If his job were taken by a second welder, the second vacancy would be filled within the department. But the vacancy created by the second move would then be posted plantwide and presumably a foundry worker or other worker with qualifications could transfer across departmental lines. This change aides Negroes to transfer out of the foundry. It came about because of pressures, non-racial, on the part of older workers who were laid off, while others, with less plant but more departmental seniority were working.

Even where plantwide seniority is applicable to promotions and transfers the requirement that one must be able to perform the job satisfactorily is a great deterrent. It is extremely difficult for a skilled molder in the foundry with many years of seniority to acquire at the same time sufficient skill to successfully bid on a welding, machine shop, or electrical job. And, of course, many black semiskilled operatives just do not have the educational backgrounds to meet skilled job requirements.

The Memphis Plant Situation

The union-management contract in effect since the late 1940's at Harvester's Memphis plant, but revised since 1964, has shaped the nature and extent of the integration to some degree. At the present time, contracts at Memphis and all other Harvester facilities contain the following clauses which guarantee non-discrimination in enforcement of the contract:

> *Section 1.* Neither the Company nor the Union in carrying out their obligations under this contract, shall discriminate in any manner whatsoever against any employee because of race, sex, age, as provided in

the Age Discrimination in Employment Act of 1967, political or religious affiliation, or nationality.

Section 2. The Company agrees to continue its present nondiscriminatory policy offering equal opportunities for available jobs to qualified applicants without regard for their nationality, race, sex, age, as provided in the Age Discrimination in Employment Act of 1967, political or religious affiliation, or membership in any labor or other lawful organization.[75]

From time to time prior to 1960, however, both the local union and local management violated the spirit of the contract by permitting discriminatory hiring practices. In addition, the much superior opportunities for better education and technical training available to the whites helped the whites to garner the skilled trades jobs. Consequently, Negroes were segregated into certain departments, such as the foundry, and excluded from certain other departments such as machining, welding, and skilled trades. In this situation, the fact that the seniority provisions of the contract confine one's rights thereunder to a departmental basis even though impartially enforced, has made it exceedingly difficult for Negroes to move into these skilled trades departments.

For example, in late 1968 a Negro press operator with ten years of press department and production unit seniority passed examinations qualifying him as an electrician. He previously had experience which helped him to qualify. According to the contract, however, after he had worked 60 days in the skilled trades electrical department, he would lose all the seniority that he formerly had in the production unit. This Negro press operator served as an electrician for approximately 50 days, 10 days short of the period in which he would lose his production job seniority. Fortunately, one might say, he was at this point laid off from the electrical department because of a reduction of work in the department. Consequently, he was able to return to his production job as a press operator since he still possessed his production seniority. If he had remained 10 more days in the electrical department and then received a layoff, he would have been unemployed until work picked up in the electrical department. Under such contract provisions as these, how many workers with production unit seniority can afford to transfer into the skilled trades even if they have the experience and ability?

---

[75] Correspondence from Mr. H. C. Baker, Employee Relations Manager, November 19, 1971.

Interestingly enough, this particular clause was not designed in Memphis for the purpose of discriminating against Negroes. It happens to be a common clause in UAW and other union plants in many sections of the country. Its purpose nationally has been to give the skilled trades the autonomy which they wish. In order to establish and support the wages which they demand and have received they insist that they are apprenticeable trades and that the way into these trades is generally through apprenticeship. The loss of plantwide seniority in the production unit which a worker incurs when he transfers to the skilled trades has been a reprisal commonly demanded by the less skilled production workers because they are not given free access into the skilled trades. But the effect of this clause in Memphis is to make it extremely difficult for Negroes to enter the skilled trades. In practical terms it leaves open to them only the route of apprenticeship with all its formal educational tests in which the whites with their superior education have the advantage.

There are other contractual hinderances to entering the skilled trades. Let us suppose that a Negro molder in the foundry wishes to become a welder. Let us assume that he had been fortunate enough to have had one of the few opportunities to learn welding, perhaps at night school. Even though his plantwide seniority would give him a right to begin at the bottom job in welding, he is given only 40 hours in the new position in which to achieve the expected earning rate on this piece rate job. In other words, he must be qualified and must demonstrate within 40 hours that he can perform at the minimum speed which the company expects of an experienced welder.

Not only transfers to skilled trades, but those to other departments within the production unit are blocked by the contract's departmental seniority preference clause. Thus if a Negro foundry employee tries to become a machine operator, inspector, or stock clerk outside the foundry, he finds that each job opening in such departments is offered first (for a period of 24 working hours) to those who already work in that department. The Negro foundry worker would find, for example, that his 20 years of plantwide seniority would not be as powerful as six years of seniority held by an employee in the department to which the Negro wishes to transfer. The long term employee who wishes to transfer must expect to go to the bottom of the job classification and wage rate in the department to which he transfers and may further run the risk of moving from a day shift to a night shift.

The Louisville Plant Situation

It is the division of the workers in a plant into several unions, each with its own seniority rules, which is an even greater barrier to the full integration of Negroes into all departments of the plant. In the agricultural and construction machinery industries the UAW represents the great majority of workers, but the craft unions have established separate seniority lines which restrict the movement of workers from production to craft units. Even in totally UAW plants skilled craft autonomy is so strong that worker movement from production to craft units is almost as much impeded as if the craft group belonged to a separate union. At International Harvester's Louisville tractor manufacturing plant, the two big divisions of the plant, the foundry and assembly, are split in their jurisdiction between the two UAW local unions. UAW Local Union No. 1336 has jurisdiction over what is known as the machine shop, and UAW Local No. 817 has jurisdiction over the foundry in the same plant. Each local union has completely separate seniority rosters. Years of service in one local cannot be transferred to the other. Hence, practically speaking, it is not feasible for a Negro in the foundry to transfer to the assembly or machine shop.

It should be noted that the UAW did not set up these two unions for the purpose of preserving discrimination. When the plant was built, competition between the UAW and the Farm Equipment Workers' Union resulted in a National Labor Relations Board election in which the machine shop employees voted for the Farm Equipment Workers and the foundry employees for the United Auto Workers. When in 1955 the UAW was able to win the machine shop from the Farm Equipment Workers (then the UE), it merely continued the old Farm Equipment Workers local union with the same jurisdiction instead of merging the locals.

Now in this current drive for integration the union could facilitate transfer of Negroes to traditionally "white" jobs either by revising the contracts of the two Louisville locals or by merging the separate locals into one. The fact that it has not done either is a reflection of the limited power of a union with a big white majority to promote positively more rapid job integration, especially when it involves upsetting long standing seniority patterns. But surely the company and the union will have to provide greater interdepartmental mobility if they are to further integration.

Seniority—Final Comment

When examining seniority and its current relation to employment integration, it is well to recollect that the employed Negro is also a major beneficiary of seniority. Since he was the most likely victim of prejudice, union contract clauses on job bidding by seniority guaranteed him his chance to advance to the top jobs in his department solely on the basis of length of service and ability to handle the job. Enactment of seniority clauses for promotion and layoff were the result of union pressures during the late nineteen thirties and forties. They were the workers' first assault against favoritism and discrimination including race discrimination. On the whole they have been remarkably effective instruments of especial benefit to Negroes. They do not, however, solve many of the Negroes' in-plant problems. Seniority does nothing to prevent discrimination at the hiring gate nor in promotion to supervisory jobs outside the orbit of the union's legal bargaining unit.

The ending of discriminatory hiring, which the leading firms in the agricultural and construction machinery industries have done, will in the long run solve most complaints against seniority. But there is an interim period, the present, in which the results of past discrimination at the hiring gate (i.e., concentrating Negroes in the foundry, forge, etc.) linger on and, indeed, are magnified by occupational or departmental seniority. In effect, the union contract institutionalizes the status quo and inhibits change.

The companies and unions in the farm equipment and construction machinery industries are now under great pressure, along with American industries and unions in general, to alter institutions which continue in the present the effects of past discrimination. In a number of significant cases, the courts have forced the alteration of seniority in order to permit greater opportunities for advancement and upgrading for blacks who were employed under discriminatory patterns.[76] Although these cases

[76] *Quarles* v. *Philip Morris, Inc.*, 279 F. Supp. 505 (E.D. Va., 1968); *United States* v. *Local 189, United Papermakers and Paperworkers, et al.*, 282 F. Supp. 39 (E.D. La., 1968); affirmed, 416 F.2d 980 (5th Cir. 1969); cert. denied, 397 U.S. 919 (1970).

originated in situations such as the tobacco and paper industries where the discrimination perpetuated was obviously intended to be invidious,[77] the principle, known as the "rightful place" doctrine, has also been applied to industries in which the situation is analogous to that described here.[78] In effect the remedy enforced by the courts pursuant to Title VII of the Civil Rights Act of 1964 permits the "affected class"—that is those Negroes who were employed prior to the elimination of discriminatory hiring —to advance on the basis of plantwide instead of occupational or departmental seniority. Of course, the person so advancing must be qualified to do the work. Such a remedy could conceivably be applied to some plants in the farm equipment and construction machinery industries.

Any revision of seniority which would retroactively aim at correcting past discrimination against Negroes at the hiring gate by wiping out long accumulated seniority rights of other workers will set worker against worker. Meanwhile, unions and union members are generally cooperating with employers and government in the broad distribution of new employees from minority groups into all types of occupations and departments throughout the plant. This does not threaten existing "job property" rights.

GOVERNMENT POLICY IMPACT

All companies in the two industries studied are required by Title VII of the Civil Rights Act of 1964 to employ persons without discrimination because of race, color, creed, or sex. This Act also established the Equal Employment Opportunity Commission. Many states have similar laws. Judging from the relative lack of litigation involving these industries, it is likely that overt discrimination, at least in the bigger firms, has not been a serious problem since Title VII became effective.

On the other hand, this does not mean that the Equal Employment Opportunity Commission is satisfied with the progress of the industry. Wherever, as in the case of farm equipment and

---

[77] For a discussion of the situation of these industries, see Northrup, Rowan, *et al., Negro Employment in Southern Industry, op. cit.*, Part One, pp. 39-42, 95-104; and Part Three, pp. 33-37, 64-87.

[78] *United States* v. *Roadway Express, Inc.*, Civil Action No. C-68-321 (D.C. Ohio); 63 CCH Lab. Cases, #9516 (1970); *United States* v. *Bethlehem Steel Corp., Local 2602 United Steelworkers, et al.*, U. S. Court of Appeals, 2nd Cir., June 21, 1971.

construction machinery, the percentage of minority employees is below that of industry generally or manufacturing industry or below the black population of the area in which a plant is located, the EEOC is pointing out these shortcomings to industry. The "gap in understanding" that Professor Bernard E. Anderson noted between the management in public utilities and the EEOC is equally applicable in the farm equipment and construction machinery industries:

> The commission, being an enforcement agency, looks to the statistical record for evidence that equal employment, in fact, exists in business firms. Corporate managers, however, often emphasize company progress in changing the focus, direction, and energy devoted to equal employment issues. Such changes within the business firm may not result in significant numerical gains in the number of black workers employed or promoted during a one- or two-year period. The modification of traditional employment practices is no less real to corporate managers even though the results may not meet EEOC measurement criteria. To management, changes in the employment process that might result in long term gains for black workers are often considered more important than short term gains in numbers employed. The difference of opinion between EEOC and industry representatives regarding the meaning and relative importance of affirmative action programs and affirmative action results is one of the most difficult issues in the equal employment area today.[79]

Management, on the other hand, is faced with certain difficulties such as the inertia if not opposition to employment integration among its employees. Even when this is overcome, management is often unable to make rapid changes. For example, it cannot fire current white employees in order to increase the black percentage. During years of employment decline in the industries it is very difficult to make significant gains in the percentage of black employees. Also, in the job categories in which EEOC is most insistent, clerical and technical, industry faces a bottleneck in supply.

A second government action, the executive orders establishing standards on minority employment for firms receiving government contracts, has had a more significant impact than the EEOC. Especially important for the industries are the rules and regulations adopted pursuant to Executive Order No. 11246, particularly General Order No. 4 of the Office of Federal Contract

[79] Bernard E. Anderson, *Negro Employment in Public Utilities*, Studies of Negro Employment, Vol. III (Philadelphia: Industrial Research Unit, Wharton School of Finance and Commerce, University of Pennsylvania, 1970), p. 207.

Compliance. The Executive Order sets standards above and beyond Title VII which must be adhered to by government contractors. General Order No. 4 issued on November 20, 1969 sets forth these standards in great detail. They include: 1) an analysis of all major job classifications and an explanation of why minorities may be underutilized; 2) the establishment of goals, targets and affirmative action commitments designed to relieve any shortcoming identified; and 3) the development and supply of data to government organizations, including not only racial-occupational figures in great detail, but also progression charts, seniority rosters, and an analysis of applicants by race for the various jobs.

To be granted a federal contract or subcontract and to retain a contract already granted, a company must be judged "in compliance" with these standards. The judgment of "in compliance" or "non-compliance" is based upon company reports and plant inspections by government inspectors. Since federal government contracts are granted by agencies such as the Defense Department and the General Services Administration, these contracting agencies, as well as the OFCC, have their own equal oportunity staffs and make company inspections.

Where inspection reveals that a company is not meeting the OFCC standards it may be required to take certain affirmative actions in order to gain its "in compliance" rating. Such actions may include recruitment visits to secondary schools and colleges with high Negro enrollment to find Negroes to fill white collar jobs, advertising for employees in Negro community newspapers, ending of recruitment restricted either to walk-ins or to referrals by families of previously employed employees, establishment of special training courses to aid in the transfer and upgrading of Negro employees into all-white departments, special in-plant recruitment to increase the proportion of Negro foremen and programs for "hardcore" recruitment and training.

Generally, companies are required to appoint an executive called an "Equal Employment Opportunity Coordinator." By making this position either a full-time position or one in which the executive has major responsibility, the government hopes to prevent the normal production, sales, or other executive duties from crowding out attention to equal employment. All the large companies in these two industries have appointed such coordinators. Sometimes there are such coordinators even in the individual plants as well as one for the entire company.

In spite of the efforts of OFCC to establish concrete standards for employment integration, the judgment of "in compliance" or "non-compliance" is, to a great extent, a subjective one. The decision is made especially difficult because of the possible dire consequences of the only overt response provided for a ruling of "non-compliance," i.e., the denial or cancellation of a government contract which may put a plant out of business and/or prevent or hinder the government from obtaining necessary armaments or other materials.

Consequently, the strictness of the application of the OFCC standards varies considerably. In some cases companies with comparatively good records may be threatened with a rating of "non-compliance" because government officials think that their rate of improvement is too slow. Other companies with serious failures in Negro employment and upgrading may be designated "in compliance" upon merely submitting "an acceptable program for future compliance."

The most important and universal finding of the government plant inspectors has been the preponderance of Negroes in the foundry and their absence in the skilled trades. In one such case, in 1963 a company's foundry which had only 25 percent of the total plant employment employed 90 percent of its total Negro employees. At the same time eight departments were listed as having no Negro employees. By 1966 this situation had not changed significantly despite several OFCC warnings and suggestions. But, by 1968 this particular company had changed its employment policy so that newly employed whites were being sent to the foundry and efforts were being made to transfer Negroes already in the foundry to other departments.

In a second company, in late 1966, one-third of companywide Negro hires were assigned to the foundry, two-thirds to the assembly line. By mid-1967 the assignment of new Negro employees was as follows: to the foundry, one; to the assembly, 74; to the forge, 2; to plant protection, 3.

Of course, in some areas such as sales, the government impact has been virtually nil. A particular company was criticized because, although it appeared to be making progress in integrating its work force, it had made absolutely no progress in putting Negroes into sales positions. However, in spite of such criticism and counseling for special efforts to employ Negro salesmen, the data do not as yet show an increase in this occupation.

The inspections do have an educational effect on management thinking. In one case a company which had made an outstanding effort to improve its employment integration was sharply criticized because during a conference with the agency inspector, a superintendent had referred to some of the Negro employees as "boys." In this case the company officials expressed appreciation for this correction and felt that the viewpoint of the agency inspector, in this case a Negro, had been very helpful.

There can be no doubt that these government pressures have become a most significant factor in augmenting the progress of black employment. They are particularly significant in an industry such as farm equipment in which employment is not expanding.

CHAPTER VI

The Determinants of Industry Policy

Throughout this study we have discussed several factors which have affected black employment and upgrading in the farm equipment and construction machinery industries. By way of a conclusion, we shall now summarize these factors.

DEMAND FOR LABOR

Black employment has increased during periods of general employment expansion and has reflected to an exaggerated degree the contractions of the labor market. The percentage of Negroes employed increased during the labor shortage of World War II. At this time, however, the expansion was predominately in the blue collar area. Virtually no progress was made in the skilled trades, in supervisory jobs in the plants, or in white collar jobs in the offices. Progress in Negro employment lagged in the economic recession of the late 1950's and early 1960's, but in the industrial growth of the mid-1960's Negroes again increased their proportion of jobs in these two industries. In the last years of this decade employment declined in the farm equipment industry and Negroes suffered disproportionately in the layoffs, not only because they were generally the last hired, but also because they were overconcentrated in the areas most affected by the decline, i.e., semiskilled and unskilled blue collar occupations.

Thus it appears that the most fundamental factor in achieving truly equal employment for blacks is expanding general employment. In nineteenth century America the land hunger and poverty of the poor and the immigrants could be alleviated by offering them the limitless government-owned western lands. Thus, the United States was spared the ideological and class strife which usually accompanies attempts to redistribute land already privately owned. Likewise, today equal employment for blacks can be accomplished most completely and peaceably if there are new jobs to be filled instead of solely existing jobs to be redivided between whites and blacks.

JOB STRUCTURE

The job structure of our two industries is favorable for black employment. Almost 70 percent of the jobs are blue collar and about 45 percent of the jobs are semiskilled or unskilled. Given the fact that Negroes are disproportionately underrepresented in salaried and skilled positions, a job structure of the nature found here can result in a high black labor utilization if other factors are also favorable. Locational factors, however, offset this job structure advantage to a considerable degree.

LOCATIONAL FACTORS

There seems little doubt that locational factors are a fundamental cause of the relatively low participation of Negroes in these two industries. Where the plants are concentrated, Negro population is low; and the situation has worsened in recent years because of the closing of several Chicago facilities of the International Harvester Company. Moreover, where the plants are located near large cities, they tend to be a considerable distance from the areas where most blacks live. As long as current population and housing trends continue, and unless areas in which Negroes are concentrated can be made more suitable for industrial plant locations, it is unlikely that the farm equipment and construction machinery industries will become large employers of Negroes.

MANAGEMENT POLICIES

In the final analysis, equal employment requires managerial action. Whatever the pressures and wherever the problems, it is management that must act. An outstanding example of such action and commitment was provided by International Harvester which devised and implemented an equal employment program long before such action was required by law. Later the leaders of other large companies in these industries accepted the challenge of doing more than merely not discriminating. Obviously, the increases in black employment which occurred in the 1960's were not just the results of market forces. Rather they reflected the impact of committed companies, acting under government pressure, taking affirmative action to find and to employ minorities in greater numbers and proportions.

The data in this report indicate that in spite of management efforts job upgrading for Negroes is proceeding slowly in the large companies and has hardly begun in the small ones. One limiting factor has been the lack of trained applicants. Of course, management's own training programs cannot be expected to counter the massive failure of our society to give equal educational opportunity to blacks.

It is interesting to note that neither private customer or consumer pressure is a significant factor affecting minority employment in these industries. It would be most unlikely that sales to farmers or to the construction industry would be a likely vehicle with which to pressure companies to expand such employment. Management policies and programs, spurred by government pressure, operate without the pressure of the buying public found significant in some other industries.

UNION POLICIES

In manufacturing hiring has been the exclusive prerogative of management. Today, management in these two industries is working to undo the results of the discriminatory hiring policies of earlier years by upgrading their black employees. The unions in the industry have played a more passive role in this effort. Local unions have cooperated with management (sometimes in response to prodding by their national officers) in placing nondiscrimination clauses in the labor contracts. They have cooperated in finding and aiding black candidates for special, hard-core job training programs. However, no local union visited by the author was proposing to redress past job discrimination against blacks by abolishing or even overlooking departmental or occupational seniority. A democratically elected leadership of a local union with a majority of white members can act less decisively than can an appointed management.

The vigorous promotion by national union leaders of nondiscrimination in the local unions and equal employment in the plants is an important force as seen in the United Automobile

Workers International takeover of a formerly racist-behaving local union at the International Harvester Memphis plant.

GOVERNMENT POLICIES

Government policies have been a powerful inducement for progress in Negro employment in the two industries. They not only require action of those not inclined to alter the status quo, but they also benefit those managements which have for some years demonstrated genuine committment to equal opportunity. Time and again government inspections have pinpointed gaps in management performance: lack of Negroes in sales, clerical, and skilled trades; even segregated facilities in the South. Under government counsel management is learning new techniques of job recruitment and training and is re-examining its testing procedures for job promotion. In addition the government equal employment standards help protect the more progressive employers and unions against overt reaction by prejudiced employees.

Thus far government has not dealt with the plant location problem which is limiting seriously Negro employment in these two industries. This problem of bringing together the unemployed blacks of the ghettos and the new plants located in the suburbs and rural areas is one needing government study and action. Of course, this will involve the breaking of segregated housing paterns and the study of the possibilities of again locating manufacturing plants within the city.

A government responsibility of paramount importance in achieving equal employment is the maintenance of a steadily expanding economy in order that Negroes can find job opportunities more readily. In addition, greater training and improved educational opportunities are necessary if blacks are to achieve equal employment in crafts, white collar, and professional categories. Only government has the financial resources to provide this education and training for all blacks. Thus, we conclude here, as has been concluded in other studies in this series, that government pressure is a necessary ingredient for continued black employment progress.

CONCLUDING REMARKS

Unless the newly located plants and the Negro populations can be brought together, neither the farm equpiment nor the construction machine industry is likely to be a leading employer of black labor during this decade. The construction machinery industry, by reason of its greater growth and expansion, will probably employ both a larger number and a higher proportion of Negroes than will farm equipment. In both industries, the proportion of Negroes is likely to increase as a result of management policy and government pressure. The extent of this increase will depend greatly upon community and national policies in plant location, housing, and education.

Undoubtedly an expanding economy and accompanying strong labor demand will be the prime factors in bringing blacks into the work force in greater numbers and at higher levels. This strong demand for labor can supply important motivation to society to tackle its problems of location and training.

Index

Major Industrial Research Unit Studies

No. 51 Gordon F. Bloom, F. Marion Fletcher, and Charles R. Perry, *Negro Employment in Retail Trade: A Study of Racial Policies in the Department Store, Drugstore, and Supermarket Industries*, Studies of Negro Employment, Vol. VI. 1972. $10.00

No. 50 Herbert R. Northrup *et al.*, *Negro Employment in Land and Air Transport: A Study of Racial Policies in the Railroad, Airline, Trucking, and Urban Transit Industries*, Studies of Negro Employment, Vol. V. 1971. $13.50

No. 49 Herbert R. Northrup, Richard L. Rowan, *et al.*, *Negro Employment in Southern Industry: A Study of Racial Policies in the Paper, Lumber, Tobacco, Coal Mining, and Textile Industries*, Studies of Negro Employment, Vol. IV. 1971. $13.50

No. 48 Bernard E. Anderson, *Negro Employment in Public Utiliites: A Study of Racial Policies in the Electric Power, Gas, and Telephone Industries*, Studies of Negro Employment, Vol. III. 1970. $8.50

No. 47 Armand J. Thieblot, Jr., and Linda P. Fletcher, *Negro Employment in Finance: A Study of Racial Policies in Banking and Insurance*, Studies of Negro Employment, Vol. II. 1970. $9.50

No. 46 Herbert R. Northrup, Richard L. Rowan, *et al.*, *Negro Employment in Basic Industry: A Study of Racial Policies in Six Industries*, Studies of Negro Employment, Vol. I. 1970. $15.00

No. 45 William N. Chernish, *Coalition Bargaining: A Study of Union Tactics and Public Policy.* 1969. $7.95

No. 44 Herbert R. Northrup and Gordon R. Storhom, *Restrictive Labor Practices in the Supermarket Industry.* 1967. $7.50

No. 48 F. Marion Fletcher, *Market Restraints in the Retail Drug Industry.* 1967. $10.00

No. 42 Michael H. Moskow, *Teachers and Unions: The Applicability of Collective Bargaining to Public Education.* 1966. $8.50

No. 41 George M. Parks, *The Economics of Carpeting and Resilient Flooring: An Evaluation and Comparison.* 1966. $5.00

No. 40 Gladys L. Palmer *et al.*, *The Reluctant Job Changer.* 1962. $7.50

Labor Relations and Public Policy Series

3. *The NLRB and the Appropriate Bargaining Unit*, by John E. Abodeely. 1971. $5.95
4. *The NLRB and Secondary Boycotts*, by Ralph M. Dereshinsky. 1972. $5.95
5. *Collective Bargaining: Survival in the '70's?* Conference Proceedings, edited by Richard L. Rowan. 1972. $7.95
6. *Welfare and Strikes: The Use of Public Funds to Support Strikers*, by Armand J. Thieblot, Jr., and Ronald M. Cowin. 1972. $6.95

Miscellaneous Series

17. *The Carpet Industry: Present Status and Future Prospects*, by Robert W. Kirk. 1970. $5.95
18. *Educating the Employed Disadvantaged for Upgrading: A Report on Remedial Education Programs in the Paper Industry*, by Richard L. Rowan and Herbert R. Northrup. 1972. $5.95

Racial Policies of American Industry Series

1.	*The Negro in the Automobile Industry,* by Herbert R. Northrup.	1968.	$2.50
2.	*The Negro in the Aerospace Industry,* by Herbert R. Northrup.	1968.	$2.50
3.	*The Negro in the Steel Industry,* by Richard L. Rowan.	1968.	$3.50
4.	*The Negro in the Hotel Industry,* by Edward C. Koziara and Karen S. Koziara.	1968.	$2.50
5.	*The Negro in the Petroleum Industry,* by Carl B. King and Howard W. Risher, Jr.	1969.	$3.50
6.	*The Negro in the Rubber Tire Industry,* by Herbert R. Northrup and Alan B. Batchelder.	1969.	$3.50
7.	*The Negro in the Chemical Industry,* by William Howard Quay, Jr.	1969.	$3.50
8.	*The Negro in the Paper Industry,* by Herbert R. Northrup.	1969.	$8.50
9.	*The Negro in the Banking Industry,* by Armand J. Thieblot, Jr.	1970.	$5.95
10.	*The Negro in the Public Utility Industries,* by Bernard E. Anderson.	1970.	$5.95
11.	*The Negro in the Insurance Industry,* by Linda P. Fletcher.	1970.	$5.95
12.	*The Negro in the Meat Industry,* by Walter A. Fogel.	1970.	$4.50
13.	*The Negro in the Tobacco Industry,* by Herbert R. Northrup.	1970.	$4.50
14.	*The Negro in the Bituminous Coal Mining Industry,* by Darold T. Barnum.	1970.	$4.50
15.	*The Negro in the Trucking Industry,* by Richard D. Leone.	1970.	$4.50
16.	*The Negro in the Railroad Industry,* by Howard W. Risher, Jr.	1971.	$5.95
17.	*The Negro in the Shipbuilding Industry,* by Lester Rubin.	1970.	$5.95
18.	*The Negro in the Urban Transit Industry,* by Philip W. Jeffress.	1970.	$4.50
19.	*The Negro in the Lumber Industry,* by John C. Howard.	1970.	$4.50
20.	*The Negro in the Textile Industry,* by Richard L. Rowan.	1970.	$5.95
21.	*The Negro in the Drug Manufacturing Industry,* by F. Marion Fletcher.	1970.	$5.95
22.	*The Negro in the Department Store Industry,* by Charles R. Perry.	1971.	$5.95
23.	*The Negro in the Air Transport Industry,* by Herbert R. Northrup and others.	1971.	$5.95
24.	*The Negro in the Drugstore Industry,* by F. Marion Fletcher.	1971.	$5.95
25.	*The Negro in the Supermarket Industry,* by Gordon F. Bloom and F. Marion Fletcher.	1972.	$5.95
26.	*The Negro in the Farm Equipment and Construction Machinery Industries,* by Robert Ozanne.	1972.	$5.95
27.	*The Negro in the Electrical Manufacturing Industry,* by Theodore V. Purcell and Daniel P. Mulvey.	1971.	$5.95
28.	*The Negro in the Furniture Industry,* by William E. Fulmer.	1972.	$5.95
29.	*The Negro in the Longshore Industry,* by Lester Rubin.	1972.	$5.95
30.	*The Negro in the Offshore Maritime Industry,* by Lester Rubin and others.	1972.	$5.95

Order from the Industrial Resea
4025 Chestnut Street, Philadelphia, Pen